Lead-Free Soldering

Lead-Free Soldering

Edited by

Jasbir Bath
Solectron Corporation

 Springer

Jasbir Bath
Solectron Corporation
637 Gibraltar Court, Building 1
Milpitas, CA 95035
USA

Library of Congress Control Number: 2007920042

ISBN 978-0-387-32466-1   e-ISBN 978-0-387-68422-2

Printed on acid-free paper.

9 8 7 6 5 4 3 2 1

springer.com

Preface

The past few years have seen major developments in soldering materials and processes for electronics assembly manufacture due to the movement from tin-lead to lead-free soldering. The removal of lead from electronics solders due to environmental considerations first developed with proposed US legislation in the early 1990s. At that time, the alternatives had not been fully explored, so a ban on the use of lead in electronic solders was put on hold. However the seed was sown for development with various projects initiated during the 1990s in Europe, the Americas, and Asia.

Based on government pressures, Japan OEMs began to move to lead-free solder products from 1998 and this, combined with the European Union ROHS (Restriction of Hazardous Substances) legislation enacted in 2006, drove the global manufacture of electronics consumer products with lead-free solders. From 1998 to the present, the development of lead-free solder materials and processes has progressed to such an extent that development work moving forward will typically only concentrate on lead-free solders and components rather than tin-lead solders and components.

This book aims to give the latest information on development of the lead-free soldering materials and processes and identify where more work is needed. The chapters of the book describe legislation, alloys, reflow, wave, rework, reliability, backward and forward process compatibility, PCB surface finishes and PCB laminates, and standards affecting the general lead-free soldering arena.

The information in the book is provided by many authors who are fully immersed in the transition from tin-lead to lead-free soldering as part of their daily work lives. It is our hope that this book provides a useful source of knowledge and information for process engineers and other functions and stimulates further work in this area.

The chapter authors are to be thanked for spending their time and effort to create their respective chapters, which is difficult to do when combined with a busy work and personal family schedule.

Jasbir S. Bath
Solectron Corporation
January 2007

Table of Contents

List of Authors

Jennifer Shepherd
Canyon Snow Consulting LLC
459 Monterey Avenue
Los Gatos
CA 95030
USA
E-mail: jennifershepherd@canyonsnow.com

Carol Handwerker
Purdue University
Materials and Electrical Engineering Building
501 Northwestern Avenue
West Lafayette
IN 47907-2044
USA
E-mail: handwerker@purdue.edu

Kil-Won Moon
NIST (National Institute of Standards and Technology)
100 Bureau Drive, Stop 8555
Gaithersburg
MD 20899-8555
USA
E-mail: Kil-Won.Moon@nist.gov

Ursula Kattner
NIST (National Institute of Standards and Technology)
100 Bureau Drive, Stop 8555
Gaithersburg
MD 20899-8555
USA
E-mail: Ursula.Kattner@nist.gov

Sundar Sethuraman
Solectron Corporation
637 Gibraltar Court, Building 1
Milpitas
CA 95035
USA
E-mail: sundarsethuraman@ca.slr.com

Christiane Faure
Solectron Corporation
109E Chemin Departemental
CANEJAN BP 6
CESTAS CEDEX 33611
Bordeaux
France
E-mail: christianefaure@fr.slr.com

Jasbir Bath
Solectron Corporation
637 Gibraltar Court, Building 1
Milpitas
CA 95035
USA
E-mail: jasbirbath@ca.slr.com or jasbir_bath@yahoo.com

Jianbiao (John) Pan
Cal Poly State University
Dept. of Industrial & Manufacturing Engineering
San Luis Obispo
CA 93407
USA
E-mail: pan@calpoly.edu

Dennis Willie
Solectron Corporation
637 Gibraltar Court, Building 1
Milpitas
CA 95035
USA
E-mail: denniswillie@ca.slr.com

Xiang Zhou
Solectron Corporation
637 Gibraltar Court, Building 1
Milpitas
CA 95035
USA
E-mail: XiangZhou@ca.slr.com

Jean-Paul Clech
For regular mail:
EPSI Inc.
P.O. Box 1522
Montclair
NJ 07042
USA
E-mail: JPClech@aol.com

For shipping with Fedex. UPS, Airborne etc:
EPSI Inc.
Attn: J-P. Clech
101 Gates Avenue, Ste. H-10
Montclair
NJ 07042
USA
E-mail: JPClech@aol.com

Karl Sauter
Sun Microsystems, Inc.
Semiconductor Packaging & PCB Technology
4150 Network Circle
Mailstop: SCA12-306
Santa Clara
CA 95054
USA
E-mail: Karl.Sauter@Sun.COM

Hugh Roberts
Atotech USA Inc.
1750 Overview Drive
Rock Hill
SC 29730
USA
E-mail: Hugh.Roberts@atotech.com

Kuldip Johal
Atotech USA Inc.
1750 Overview Drive
Rock Hill
SC 29730
USA
E-mail: Kuldip.JOHAL@atotech.com

For my Mum (Kuldip Kaur Bath), Dad (Harbans Singh Bath), wife (Piyanoot) and three little girls (Palm, Mint and Surinder)

Introduction

The discussion to create this book on lead-free soldering came in October 2005 during a U.S. National Electronics Conference organized by SUNY-Binghamton. There appeared to be a need for a lead-free soldering book which covered subjects in detail but not to a depth which would not be practical for the process engineer. This book brings together a diverse array of persons from different companies and expertise who are currently in the area of lead-free soldering development. The contents of the book provide a guide to understanding the main issues in lead-free soldering.

Lead-free and other environmental regulatory legislation is appearing rapidly and affecting the electronics industry. Engineers need to be aware of what will affect them currently and what is potentially on the horizon to help them to adjust and adapt. The first chapter (by Canyon Snow Consulting LLC) covers a review of existing and upcoming lead and other material restriction legislation and discusses the driving forces behind them.

Although there have been many book chapters and paper discussions of lead-free alloys, there is a need for persons (from Purdue University and NIST (National Institute of Science and Technology)) who have been involved in the forefront of lead-free alloy development to review various properties of lead-free solder alloys, in particular those alloys which are being predominantly used in the industry. This book provides an in-depth review of them and a discussion of the issues surrounding them in Chapter 2.

The main use of these alloys is in surface mount assembly, and Chapter 3 reviews SMT lead-free reflow soldering which focuses on the main lead-free alloy of choice: tin-silver-copper (SnAgCu). An area where there is a minimal amount of research papers and virtually no book chapters available is lead-free wave soldering as the focus of lead-free research development has centered on surface mount assembly. A chapter on lead-free wave soldering was identified as a gap which would be of use to the process engineer (Chapter 4). Even though rework is not desired, and should be minimized, it is part of the manufacturing process. There is a minimal amount of research papers and virtually no book chapters available on lead-free rework and this was also identified as a gap to be of use to the process engineer (Chapter 5). All three of these chapters were written by authors from an electronics manufacturer (Solectron Corporation).

After the board is assembled, the main question which is asked is 'will the assembled solder joint be reliable?'. Chapter 6 concentrates on one aspect of reliability from an author (from EPSI Inc.) who has been at the forefront of lead-free reliability investigations. This chapter begins with a summary of trends in reliability test results for some of the mainstream lead-free alloys used in board assemblies. It then presents a case study with the goal of presenting the type of data, material properties and analysis that are needed to estimate component attachment reliability for a given soldering alloy under thermal cycling conditions. One objective of this exercise is to illustrate the level of details that are required to develop stress/strain analysis models and estimate component attachment reliability in lead-free board assemblies. By definition, reliability is product and application specific and blanket statements about the reliability of a lead-free assembly need to be considered with caution. Part of the solder joint assembly reliability discussions also focuses on the transition to lead-free soldering where a lead-free solder paste or wave alloy may be assembled with tin-lead components, and a tin-lead solder paste or wave alloy may be assembled with lead-free components. Although there have been some papers discussing whether there are any reliability issues with this 'mixed' assembly processing, there have been few or no book chapters reviewing this specific area. The author of this chapter (from Cal Poly State University) gives an objective and detailed discussion of the issues with possible solutions. Chapter 7 also discusses the use of lead-free press-fit components from an electronics assembler (Solectron Corporation) with first hand experience of some of the potential issues that can be faced.

One area which is a work in progress is the development of PCB laminate materials for lead-free soldering, particularly laminates for thicker high end reliability boards. The chapter author (from SUN Microsystems) gives an objective and focused review of PCB laminates for lead-free soldering with testing methodologies and a discussion of the need and reasons for higher temperature-rated laminates for lead-free soldering (Chapter 8). There have been many papers written about lead-free board finishes but there have been few book chapters which give an objective and detailed discussion of the different lead-free board surface finishes available. The authors, who are part of a chemical plating supplier company (Atotech Inc.), provide a clear discussion of the subject which will be a good reference guide for process engineers (Chapter 9). The various advantages and disadvantages of the different lead-free board surface finishes are discussed.

No book on lead-free soldering should typically be finished without a review of existing and developing lead-free standards identifying where there are gaps in development. Chapter 10 is not a complete list of existing and developing standards, but of those which have or will likely have a role to play as reference standards for process engineers and is written by an electronics assembly manufacturer (Solectron Corporation).

Although each chapter has its own conclusion, a small conclusions section at the end of this book covers a discussion of the main conclusions from each chapter, with a discussion of some of the areas which will need to be explored and developed in the future. It is our hope that this book will provide a good grounding of the general areas for lead-free soldering for process engineers and other functions and stimulate discussion and development in this area.

Chapter 1: Lead Restrictions and Other Regulatory Influences on the Electronics Industry

Jennifer Shepherd, Canyon Snow Consulting LLC

1.1 Introduction

As technologists and engineers prepare for and address important changes to chemical content of electronic products, it would be useful to understand the drivers, context, and trends for this activity. This introductory chapter will provide a brief overview of environmental legislative and regulatory trends that are influencing the movement to lead-free electronics and will attempt to set the stage for thinking about future challenges.

1.1.1 Historic Context: What Existed Before?

Those familiar with environmental compliance requirements placed on industrial operations in North America and other regions over the last several decades will recall that the vast majority of scrutiny focused on the operation of facilities and the chemicals used in manufacturing processes. Industrial plants were governed by the laws of the country, state, or region in which they were situated and certain limits and restrictions were imposed on activities like discharging wastewater, emitting air pollutants, and disposing or recycling hazardous and solid waste.

Those days have been replaced in favor of a new global focus on the end result of manufacturing processes: the actual product. While the original requirements on facilities remain in effect, a new set of expectations has now been placed on products and the environmental impacts of those products over the entire course of their useful lives.

1.1.2 Forces for Change: What Are the Drivers?

This new focus on the product is part of a greater movement to incorporate an integrated product policy (IPP), driven primarily by European legislators. The aim is to increase or extend the responsibility of producers (commonly referred to as EPR, or Extended Producer Responsibility) who place products on the market for use by consumers and industrial end-users. Some legislators are expressing a belief that the total environmental cost of the product and ownership must be taken into account before improvements to environmental impacts can be made.

Although legislation like the Restriction of Hazardous Substances (RoHS) [1] officially affects products placed on the market in the European Union only, the realities of the global marketplace are such that technical requirements for one region have translated to additional markets. Producers find it uneconomical to produce a RoHS-compliant product for the EU market while continuing to produce non-RoHS-compliant products for the rest of the world markets. They have thus made RoHS compliance a goal for all products.

As we will discuss later, similar legislation is emerging in other jurisdictions and markets, increasing the number and complexity of regulatory burdens placed on the design of electronic products.

1.1.3 Current Trends: Where Are We Going?

Since environmental regulation of products in addition to production processes was first introduced in the EU, other governments have begun to initiate similar efforts. The concept of EPR has found resonance in terms of product content restrictions, end-of-life responsibilities, and eco-design requirements in Asia and the Americas as well.

There has been both a geographic spread of substance restriction legislation in electronic products, as well as the development of other types of environmental regulations that affect the design, manufacture, and marketing of these products.

1.2 A Regulatory Tour of Europe

As the birthplace and home of RoHS, the European Union is a trendsetter and an important region to monitor for hints about the future of environmental regulation of electronic products. As we have seen, there are complexities inherent in the RoHS and other legislation in effect and pending in the EU that have made implementation difficult and challenging.

Beyond RoHS, other types of legislation are on the books or in the legislative process in Europe that have implications for global environmental compliance of electronic products. Specific laws, trends, and approaches are highlighted.

1.2.1 Restriction of Hazardous Substances

The RoHS is easily the most important piece of legislation affecting electronic products today, especially as it relates to engineering and supply chain impacts. The law requires that certain products placed on the market after July 1, 2006 must not contain certain substances in excess of the allowable amount. Table 1.1 summarizes these substances and the allowable levels of each.

Table 1.1. Maximum concentration values for RoHS-restricted substances

Restricted substance	Maximum Concentration Value (MCV)
Lead and its compounds	0.1% by weight
Mercury and its compounds	0.1% by weight
Hexavalent chromium and its compounds	0.1% by weight
Cadmium and its compounds	0.01% by weight
Polybrominated biphenyls (PBBs)	0.1% by weight
Polybrominated diphenyl ethers (PBDEs)	0.1% by weight

The concept of the MCV and its point of measurement are very important to compliance with RoHS. Guidance from the EU Commission states that the MCV applies at the homogeneous level, and further defines this level to be that of a single material that cannot be mechanically separated from another material. Since mechanical separation can entail numerous methods, the resulting interpretation is that the MCV must be applied for each different type of material in each component and assembly. This rigorous definition has made invalid most attempts by producers to consider the component as the basis for compliance assessment, and has driven the requirement for information and compliance assessment to a level of greater detail.

The RoHS Directive applies to eight of the ten categories defining electrical and electronic equipment, as defined below. In numeric order, Categories

8 and 9 are exempt from RoHS, but listed later in this chapter in reference to the WEEE (Waste Electrical and Electronic Equipment) Directive [2].

The eight categories to which the RoHS Directive applies are listed below:

1. Large household appliances
2. Small household appliances
3. IT and telecommunications equipment
4. Consumer equipment
5. Lighting equipment
6. Electrical and electronic tools (with the exception of large scale stationary industrial tools)
7. Toys, leisure and sports equipment
10. Automatic dispensers (note: Categories 8 and 9 are not applicable under RoHS)

Several exemptions to the RoHS Directive were introduced in the initial version of the adopted legislation. Since the initial publication of the RoHS Directive in the Official Journal, several additional exemptions have been adopted by the TAC (Technical Adaptation Committee) and amended into the Directive. These are listed below, and represent a complete list of approved exemptions to Article 4(1) of RoHS as of the writing of this chapter:

ANNEX I
Applications of lead, mercury, cadmium, hexavalent chromium, polybrominated biphenyls (PBB) or polybrominated diphenyl ethers (PBDE) which are exempted from the requirements of Article 4(1)

1. Mercury in compact fluorescent lamps not exceeding 5 mg per lamp.
2. Mercury in straight fluorescent lamps for general purposes not exceeding:
   — halophosphate 10 mg
   — triphosphate with normal lifetime 5 mg
   — triphosphate with long lifetime 8 mg.
3. Mercury in straight fluorescent lamps for special purposes.
4. Mercury in other lamps not specifically mentioned in this Annex.
5. Lead in glass of cathode ray tubes, electronic components and fluorescent tubes.
6. Lead as an alloying element in steel containing up to 0,35% lead by weight, aluminium containing up to 0,4% lead by weight and as a copper alloy containing up to 4% lead by weight.

7. Lead in high melting temperature type solders (i.e. lead-based alloys containing 85% by weight or more lead), lead in solders for servers, storage and storage array systems, network infrastructure equipment for switching, signalling, transmission as well as network management for telecommunications, lead in electronic ceramic parts (e.g. piezo-electronic devices).

8. Cadmium and its compounds in electrical contacts and cadmium plating except for applications banned under Directive 91/338/EEC (1) amending Directive 76/769/EEC (2) relating to restrictions on the marketing and use of certain dangerous substances and preparations.

9. Hexavalent chromium as an anti-corrosion of the carbon steel cooling system in absorption refrigerators.

9a. DecaBDE in polymeric applications.

9b. Lead in lead-bronze bearing shells and bushes.

10. Within the procedure referred to in Article 7(2), the Commission shall evaluate the applications for:
    — Deca BDE,
    — mercury in straight fluorescent lamps for special purposes,
    — lead in solders for servers, storage and storage array systems, network infrastructure equipment for switching, signalling, transmission as well as network management for telecommunications (with a view to setting a specific time limit for this exemption), and
    — light bulbs,
    as a matter of priority in order to establish as soon as possible whether these items are to be amended accordingly.

11. Lead used in compliant pin connector systems.

12. Lead as a coating material for the thermal conduction module c-ring.

13. Lead and cadmium in optical and filter glass.

14. Lead in solders consisting of more than two elements for the connection between the pins and the package of microprocessors with a lead content of more than 80% and less than 85% by weight.

15. Lead in solders to complete a viable electrical connection between semiconductor die and carrier within integrated circuit Flip Chip packages.

16. Lead in linear incandescent lamps with silicate coated tubes.

17. Lead halide as radiant agent in High Intensity Discharge (HID) lamps used for professional reprography applications.

18. Lead as an activator in the fluorescent powder (1% lead by weight or less) of discharge lamps when used as sun tanning lamps containing phosphors such as BSP ($BaSi_2O_5:Pb$) as well as when used as speciality lamps for diazoprinting reprography, lithography, insect traps,

photochemical and curing processes containing phosphors such as SMS ((Sr,Ba)2MgSi2O7:Pb).

19. Lead with PbBiSn-Hg and PbInSn-Hg in specific compositions as main amalgam and with PbSn-Hg as auxiliary amalgam in very compact Energy Saving Lamps (ESL).
20. Lead oxide in glass used for bonding front and rear substrates of flat fluorescent lamps used for Liquid Crystal Displays (LCD).
21. Lead and cadmium in printing inks for the application of enamels on borosilicate glass.
22. Lead as impurity in RIG (rare earth iron garnet) Faraday rotators used for fibre optic communications systems.
23. Lead in finishes of fine pitch components other than connectors with a pitch of 0.65 mm or less with NiFe lead frames and lead in finishes of fine pitch components other than connectors with a pitch of 0.65 mm or less with copper lead-frames.
24. Lead in solders for the soldering to machined through hole discoidal and planar array ceramic multi-layer capacitors.
25. Lead oxide in plasma display panels (PDP) and surface conduction electron emitter displays (SED) used in structural elements; notably in the front and rear glass dielectric layer, the bus electrode, the black stripe, the address electrode, the barrier ribs, the seal frit and frit ring as well as in print pastes.
26. Lead oxide in the glass envelope of Black Light Blue (BLB) lamps.
27. Lead alloys as solder for transducers used in high-powered (designated to operate for several hours at acoustic power levels of 125 dB SPL and above) loudspeakers.
28. Hexavalent chromium in corrosive preventive coatings of unpainted metal sheetings and fasteners used for corrosion protection and Electromagnetic Interference Shielding in equipment falling under category three of Directive 2002/96/EC (IT and telecommunications equipment). Exemption granted until 1 July 2007.
29. Lead bound in crystal glass as defined in Annex I (Categories 1, 2, 3 and 4) of Council Directive 69/493/EEC OJ L 326, 29.12.1969, p. 36. Directive as last amended by 2003 Act of Accession.

1.2.2 Waste Electrical and Electronic Equipment (WEEE)

The Waste Electrical and Electronic Equipment (WEEE) Directive was originally drafted as part of the RoHS Directive, however, these were separated shortly afterward and each has a different basis in law. As an environmental protection law, the WEEE may be implemented more stringently in each EU Member State while the RoHS must be implemented in a uniform manner. The WEEE Directive focuses on the end-of-life treatment and recycling of electronic products and the financing of these activities by producers and end-users. Despite this main focus, the WEEE Directive has implications for product design and manufacture.

The full list of ten categories to which the WEEE Directive applies is given below:

1. Large household appliances
2. Small household appliances
3. IT and telecommunications equipment
4. Consumer equipment
5. Lighting equipment
6. Electrical and electronic tools (with the exception of largescale stationary industrial tools)
7. Toys, leisure and sports equipment
8. Medical devices (with the exception of all implanted and infected products)
9. Monitoring and control instruments
10. Automatic dispensers

A key piece of the legislation is contained in Annex II of the Directive, which concerns substances and devices that may cause challenges or introduce contamination during a normal recycling operation. The presence of these items must therefore be notified to recyclers so that they may be separated and removed prior to processing. It is important to note that Annex II does not represent a list of banned substances, but rather those which may be used if indicated properly to recyclers.

Annex II was due for revision in the spring of 2006 to reflect advances in understanding of electronic product design and of recycling capabilities, but was not updated at that time. The Technical Adaptation Committee (TAC), charged with implementation aspects of WEEE and RoHS, is currently working on this update. As of the time of writing of this chapter, the current version of Annex II is the original, as excerpted below.

ANNEX II
Selective treatment for materials and components of waste electrical and electronic equipment in accordance with Article 6(1)

1. As a minimum the following substances, preparations and components have to be removed from any separately collected WEEE:

— polychlorinated biphenyls (PCB) containing capacitors in accordance withCouncil Directive 96/59/EC of 16 September 1996 on the disposal of polychlorinated biphenyls and polychlorinated terphenyls (PCB/PCT) (1),

— mercury containing components, such as switches or backlighting lamps,

— batteries,

— printed circuit boards of mobile phones generally, and of other devices if the surface of the printed circuit board is greater than 10 square centimetres,

— toner cartridges, liquid and pasty, as well as colour toner,

— plastic containing brominated flame retardants,

— asbestos waste and components which contain asbestos,

— cathode ray tubes,

— chlorofluorocarbons (CFC), hydrochlorofluorocarbons (HCFC) or hydrofluorocarbons (HFC), hydrocarbons (HC)

— gas discharge lamps,

— liquid crystal displays (together with their casing where appropriate) of a surface greater than 100 square centimetres and all those backlighted with gas discharge lamps,

— external electric cables,

— components containing refractory ceramic fibres as described in Commission Directive 97/69/EC of 5 December 1997 adapting to technical progress Council Directive 67/548/EEC relating to the classification, packaging and labelling of dangerous substances (2),

— components containing radioactive substances with the exception of components that are below the exemption thresholds set in Article 3 of and Annex I to Council Directive 96/29/Euratom of 13 May 1996 laying down basic safety standards for the protection of the health of workers and the general public against the dangers arising from ionising radiation (3),

— electrolyte capacitors containing substances of concern (height > 25 mm, diameter > 25 mm or proportionately similar volume)

These substances, preparations and components shall be disposed of or recovered in compliance with Article 4 of Council Directive 75/442/EEC.

2. The following components of WEEE that is separately collected have to be treated as indicated:

— cathode ray tubes: The fluorescent coating has to be removed,

— equipment containing gases that are ozone depleting or have a global warming potential (GWP) above 15, such as those contained in foams and refrigeration circuits: the gases must be properly extracted and properly treated. Ozone-depleting gases must be treated in accordance with Regulation (EC) No 2037/2000 of the European Parliament and of the Council of 29 June 2000 on substances that deplete the ozone layer (4).

— gas discharge lamps: The mercury shall be removed.

3. Taking into account environmental considerations and the desirability of reuse and recycling, paragraphs 1 and 2 shall be applied in such a way that environmentally-sound reuse and recycling of components or whole appliances is not hindered.

4. Within the procedure referred to in Article 14(2), the Commission shall evaluate as a matter of priority whether the entries regarding:

— printed circuit boards for mobile phones, and

— liquid crystal displays are to be amended.

1.2.3 Energy-Using Products (EuP)

The Energy-using Products (EuP) Directive [3] has also recently passed into law in Europe. This directive proposes to systematically reduce the environmental burden of electronic products by introducing a series of standards related to environmentally-friendly design, or eco-design. The directive is a framework under which implementing measures may be set over time and with relative ease. Products targeted for these measures will be assessed for their relative impact on the environment, and the relative opportunity for improvement. Thus, one might expect that ubiquitous products consuming large relative amounts of energy will be ideal initial targets.

Although there is not a direct hazardous substance impact in this new legislation, engineers will need to monitor its development over time to understand and forecast the possible consequences on material selection, process engineering, and other design requirements. For example, eco-design decisions include the selection of materials by the producer and the production processes used to manufacture the product.

1.2.4 REACH

The Registration, Authorization, and Evaluation of Chemicals is also known as REACH in the European Union. While still in the approvals process, it is structured as a regulation, and not as a Directive aimed at Member State governments. This means that the final version of REACH, when passed, will be able to be translated immediately into law in each of the Member States without delay and without the issues of interpretation and confusion we have seen with both RoHS and WEEE.

The REACH entails very much what its name suggests: chemicals used in production processes and sold to industrial as well as consumer customers will need to be registered with the proper authorities, authorized prior to use, and evaluated for their ability to remain on the market as such. The proposed law applies to chemicals in many industries beyond electronic equipment, but the impact to the electronics industry will potentially be severe. The requirements of REACH are set to apply to alternate formations of a chemical and to alternate uses of the same formulation, requiring a separate approval and registration process for each. In the electronics industry, where it is often the case that low volumes of a large number of proprietary and rapidly evolving formulations are used, the REACH has the potential to slow or stop production processes.

1.3 A Regulatory Tour of Asia

As discussed at the beginning of this chapter, European environmental legislation has been observed and imitated in several important countries around the world. This section will take a closer look at Asia, and the various efforts underway to build extended producer responsibility to the electronics industry.

1.3.1 China Management Methods

In China, a version of RoHS is currently under development by the Ministry of Information Industry (MII) and is officially called "Management Methods on Controlling Pollution from Electronic Information Products". This law was officially promulgated in February of 2006 with March 1, 2007 announced as the effective implementation date for some requirements (restricted substance and recyclability disclosure, packaging description and labeling, and term of use disclosure), while the effective date for remaining requirements (related to the catalogue of RoHS restrictions) has yet to be announced.

The Chinese market is of growing importance for most electronics producers. China RoHS rules will represent a barrier to market and may not be harmonized with other regions like the EU or North America. Electronic Information Products (EIPs) are broadly defined in the law, and "non-catalogue" requirements are proposed to be in effect in less than one year. Many of the requirements are to be based on industry standards that do not yet exist and the MII has encouraged industry to provide recommendations. For example, labeling of products for China RoHS compliance must include indication of the safe use period (in years) before which any restricted substances in the product may be expected to leak from the product and potentially cause harm to the environment or to humans.

Another important and emerging requirement for compliance with China RoHS that differs from the European implementation is the emergence of a proposed certification and pre-market testing regime for products in approved analytical facilities. At the time of writing of this chapter, the final regulations for this particular aspect of implementation have not been finalized.

1.3.2 Japanese Initiatives

Japanese initiatives to date have focused on voluntary reduction of hazardous substances, and on energy efficiency for certain categories of products. For example, a group of companies formed the Japan Green Procurement Survey Standardization Initiative (JGPSSI), to develop practical steps to help industry meet EU requirements.

A Japanese government report, issued in 2005 following a series of studies, focused on information exchange related to RoHS-restricted substances. This led Japanese industry leaders to create a standardized chemical content labeling system for a limited set of products including appliances, televisions, and computers. This labeling scheme is called "J-MOSS", and it requires manufacturers to provide a label on certain appliances containing any of the RoHS-restricted substances in excess of given thresholds.

In 2005, Japan also updated its energy conservation laws to include requirements for producers to consider energy efficiency in design of products and to provide information for users to select products for purchase based on energy efficiency performance.

1.3.3 Other Countries Following Suit

South Korea has published a draft bill for waste management and recycling of electronic products and automobiles. While many details are yet to be released, there are aspects of RoHS, WEEE, and EuP included in the draft legislation. There are additional requirements for assessing the recyclability of a product, for substituting less-toxic materials when applicable, and for providing information to consumers on all of these aspects of the product. South Korea is a growing market for electronic products and for B2B (Business to Business) electronics transactions as well. It is a good example of the continuation of trends to implement Integrated Product Policy legislation in jurisdictions beyond the EU.

Taiwanese substance-restriction legislation has focused on mercury content in certain batteries. They have established limits on mercury in both individual dry batteries and those embedded in products, and require producers to submit documentation of compliance to authorities upon importation.

1.4 A Regulatory Tour of the Americas

It should be no surprise by this point to learn that similar legislative trends have made their way to North America. Previously, US environmental laws such as the Clean Air Act and Clean Water Act were frequently used as model legislation for other jurisdictions in the Americas. Trends from the EU are now typically the source of new legislation in the Americas, at national, state/provincial, and municipal levels.

1.4.1 US Federal and State Legislation

As of this writing, no US federal legislation has been passed to address electronic product substance content, recycling, or environmentally-friendly design objectives. On several occasions, bills have been proposed but have failed to be signed into law, either due to lack of political will or insufficient agreement among participants. Instead, numerous state and local policy-makers have taken the initiative to move such an agenda forward throughout the United States. California is a leading example, with Senate Bill 20 (later modified by Senate Bill 50) signed into law in 2003. This legislation addresses both recycling, through the collection of fees at the point of purchase, and hazardous substances, through a partial implementation of RoHS beginning in 2007. The scope of the California RoHS legislation is limited to Covered Electronic Devices (CEDs), defined as products with a screen size greater than 4 inches in diagonal. A bill to expand this scope to match that of the European RoHS has been authored, but has not yet passed into law in California. A minimum of 37 states have proposed or passed environmental legislation in a similar manner.

1.4.2 Canada and Mexico

Similar to the US, there has also been no national Canadian legislation. Individual provincial efforts are underway in at least six cases to date. Canadian e-waste legislation focuses on extended producer responsibility and provides for exceptions from certain requirements when WEEE is destined for recycling as treatment. Covered product vary slightly from province to province, but generally include computers (both desktop and laptop) as well as televisions and monitors. The first Canadian province to regulate e-waste was Alberta. Following suit are Saskatchewan, British Columbia, Ontario, and New Brunswick.

In Mexico, it is interesting to note a change in trend for development of environmental legislation. Once based primarily on US statutes and legislation, recent Mexican environmental regulations have been modeled after the European WEEE and RoHS Directives. There are efforts underway to study exposure to Persistent Organic Pollutants (POPs) from electronic wastes and have expanded legislation to minimize the environmental effects of packaging materials.

1.4.3 South America

Several South American countries have been actively proposing or passing e-waste legislation. For example, Costa Rica has drafted a new regulation requiring producers to provide for collection and recycling of e-waste, either individually or through an organization of producers. Mercosur (Argentina, Uruguay, Paraguay and Brazil) has approved a new requirement for producers to take back e-waste (electronic equipment, batteries, cell phones, lamps containing mercury or fluorescent tubes, and other mercury-containing EEE. Producers must also educate consumers. Finally, Argentina is considering a law that will incorporate both WEEE and RoHS. While it gives more flexibility than EU WEEE by allowing voluntary take-back systems, it is more stringent in terms of EU RoHS and does not include an exemption for the use of spare parts.

1.5 Business Impacts and Conclusions

There are many areas where environmental legislation will impact business with some areas/suggestions for a company to incorporate and consider discussed.

1.5.1 The Need for Vigilance

Producers of electronic products must remain vigilant in monitoring global environmental regulations. The diversity and number of new requirements being proposed, amended, passed into law, and implemented around the world and at different levels of government make it a challenge to remain compliant. As we have seen, non-compliance in most cases can result in exclusion from the market, fines, and even criminal prosecution. Most companies will find that the first of these potential consequences is sufficient motivation to achieve compliance.

This new approach to regulation has also created the need for a new approach to compliance. Producers have found value in cross-functional teams of regulatory experts, engineers, logisticians, lawyers, strategists, and others to interpret new requirements, assess business impacts, and build compliance into the DNA of the company's day to day operations. It is no longer sufficient to manage environmental compliance through an isolated function focused on process compliance and traditional environmental management practices.

1.5.2 Supply Chain and Distribution Channel Effects

The responses inside companies to respond to the demands of RoHS and WEEE and similar initiatives have also been mirrored in the external relationships those companies have with suppliers and distributors. At a minimum, there is a heightened need for communication up and down the supply chain to exchange information about content of restricted substances in components and sub-assemblies. There is also a basic need for producers to understand the handling of end-of life equipment by channel partners and recyclers or treatment facilities.

1.5.3 Financial and Legal Impacts

The additional diligence and communication needed to ensure regulatory compliance also require time and money. They expose producers to additional risks, which must be assessed and mitigated through contract provisions, supplier audits, and/or independent laboratory analysis. When a product or company is found to be out of compliance, many participants will likely be drawn into the process of fully assigning blame. This type of exercise can be costly, time-consuming, and damaging to existing business relationships.

Conclusions

The emerging wave of environmental regulations around the globe presents technical, administrative, and procedural compliance challenges to producers. Failure to comply with these requirements also carries significant business, financial, legal implications.

Electronic equipment producers must pay close attention to these evolving regulations and work with a variety of experts to bring all the elements of sustained compliance together.

References

1. DIRECTIVE 2002/95/EC OF THE EUROPEAN PARLIAMENT AND OF THE COUNCIL of 27 January 2003 on the Restriction Of the use of certain Hazardous Substances in electrical and electronic equipment (OJ L 37, 13.2.2003, p. 19)
2. DIRECTIVE 2002/96/EC OF THE EUROPEAN PARLIAMENT AND OF THE COUNCIL of 27 January 2003 on Waste Electrical and Electronic Equipment (WEEE) (OJ L 37, 13.2.2003, p. 24)
3. DIRECTIVE 2005/32/EC OF THE EUROPEAN PARLIAMENT AND OF THE COUNCIL of 6 July 2005 establishing a framework for the setting of ecodesign requirements for Energy-Using Products and amending Council Directive 92/57/EC and 2000/55/EC of the European Parliament and of the Council (OJ L 191, 22.7.2005, p. 29)

Chapter 2: Fundamental Properties of Pb-Free Solder Alloys

Carol Handwerker, Purdue University, West Lafayette, Indiana, USA
Ursula Kattner, National Institute of Standards and Technology (NIST),
   Gaithersburg, Maryland, USA
Kil-Won Moon, National Institute of Standards and Technology (NIST),
   Gaithersburg, Maryland, USA

2.1 Search for a Pb-Free Alternative to Sn-Pb Eutectic

The search for a global Pb-free replacement for Sn-Pb eutectic alloy has been an evolving process as the threat of a regional lead ban became a reality in July 2006. Over the twelve years from 1994 through 2006, the manufacturing, performance, and reliability criteria for Pb-free solder joints have become increasingly complex as relationships between the solder alloy, the circuit board materials and construction, and the component designs and materials have been revealed through widespread experimentation by companies, industrial consortia, and university researchers. The focus of this chapter is to examine the primary criteria used to develop the current generation of Pb-free solder alloys, the tradeoffs made between various properties once these primary criteria were satisfied, and the open questions regarding materials and processes that are as yet unanswered.

2.2 Primary Alloy Design Criteria

The primary alloy design criteria have been developed using Sn-Pb eutectic as a baseline. This approach is eminently logical: the new standard Pb-free alloy replaces Sn-Pb eutectic solder in a wide variety of board designs and microelectronics applications. For more than fifty years, printed wiring boards (PWB) and components had been designed around the behavior of Sn-Pb eutectic solder during circuit board assembly and in use for holding components to the PWBs. This behavior for Pb-free solder was quantified differently by different research groups over time, as more data on Pb-free solders became available. [1-23]

In the National Center for Manufacturing Science (NCMS) Pb-Free Solder Project completed in 1997, the pass-fail criteria for candidate alloys, as seen in Table 2.1, were created to represent practical restrictions on reflow profiles and maximum temperature, requirements in thermomechanical fatigue life and strength, wetting and oxidation of the molten solder alloy. [1-4]

Table 2.1. Pass-Fail Criteria used by the NCMS Pb-Free Solder Project [1-4]

Solder Property	Definition	Acceptable Levels
Liquidus Temperature	Temperature at which solder alloy is completely molten.	$< 225°C$
Pasty Range	Temperature difference between solidus and liquidus temperatures; temperature range where the alloy is part solid and part liquid.	$< 30°C$
Wettability	A wetting balance test assesses the force resulting when a copper wire is immersed in and wetted by a molten solder bath. A large force indicates good wetting, as does a short time to attain a wetting force of zero and a short time to attain a value of two-thirds of the maximum wetting force.	$F_{max} > 300\ \mu N$ $t_0 < 0.6$ s $t_{2/3} < 1$ s
Area of Coverage	Assesses the coverage of the solder on Cu after a typical dip test.	>85% coverage
Drossing	Assesses the amount of oxide formed in air on the surface of molten solder after a fixed time at the soldering temperature.	Qualitative scale
Thermo-mechanical Fatigue	Cycles-to-failure for a given percent failed of a test population based on a specific solder-joint and board configuration, compared to eutectic Sn-Pb.	Some percentage, usually > 50%
Coefficient of Thermal Expansion (CTE)	Thermal expansion coefficient of the solder alloy is the fraction change of length per °C temperature change. Value used for comparison was CTE of solder alloy at room temperature.	$< 2.9 \times 10^{-5}/°C$
Creep	Stress required at room temperature to cause failure in 10,000 minutes.	> 3.4 MPa
Elongation	Total percent elongation of material under uniaxial tension at room temperature.	$\gg 10\%$

By 1999, the WEEE and RoHS Directives led NEMI to establish a less quantitative, but no less restrictive, set of criteria. [15-16] The NEMI alloy was designed to:

- Have a melting point as close to Sn-Pb eutectic as possible
- Be eutectic or very close to eutectic
- Contain no more than three elements (ternary composition)
- Avoid using existing patents, if possible (for ease of implementation)
- Have the potential for reliability equal to or better than Sn-Pb eutectic.

All research groups agreed that there were no "drop-in" replacement alloys for Sn-Pb eutectic. Application of these criteria led to the NEMI choice of the Sn-Ag-Cu system, and specifically to Sn-3.9Ag-0.6Cu (± 0.2%) in the Sn-Ag-Cu (SAC) family of alloys as the most promising surface mount alloy solution.

Over the last five years, a worldwide consensus has developed that the general-purpose lead-free alloy is from the Sn-Ag-Cu ternary family. The Soldertec research organization specified a range of compositions Sn-(3.4-4.1)Ag-(0.5-0.9)Cu. The EU consortium project on Improved Design Life and Environmentally Aware Manufacturing of Electronics Assemblies by Lead-Free Soldering (IDEALS) in the European Union recommended the near eutectic alloy Sn-3.8Ag-0.7Cu. [11-14] (Note that all compositions are expressed as Sn-vX-yZ, where the "X" and "Z" are alloying elements in Sn, with the composition being "v" mass fraction*100 of element X, "y" mass fraction*100 of element "Z", and remainder being Sn; the quantity mass fraction*100 is abbreviated as 'wt.%'.) While numerous other ternary, quaternary and quinary lead-free alloys were investigated by large Japanese OEM's, the Japanese industry has moved over time toward the JEITA (Japan Electronics and Information Technology Industries Association) recommended alloy of Sn-3.0Ag-0.5Cu alloy, partly due to concerns over patent issues. (19-22)

Widespread cross-licensing of nearly all the tin-silver-copper family of solder alloys worldwide has meant that alloy selection within the SAC system is now driven primarily by overall performance in product applications and cost of Ag as an alloying agent to solder paste and bar. As is discussed below, the differences in terms of manufacturing and reliability among this full range of SAC alloys are generally believed to be small, based on available melting and reliability data. In the next sections we present the fundamental thermodynamic properties that distinguish Sn-Pb eutectic alloys from Pb-free alloys, with an emphasis on SAC alloys, and relate these differences to the microstructures that are produced when these alloys are used for circuit board assembly (Figure 2.1a-b) and to their behavior in manufacturing and in mechanical property and reliability testing.

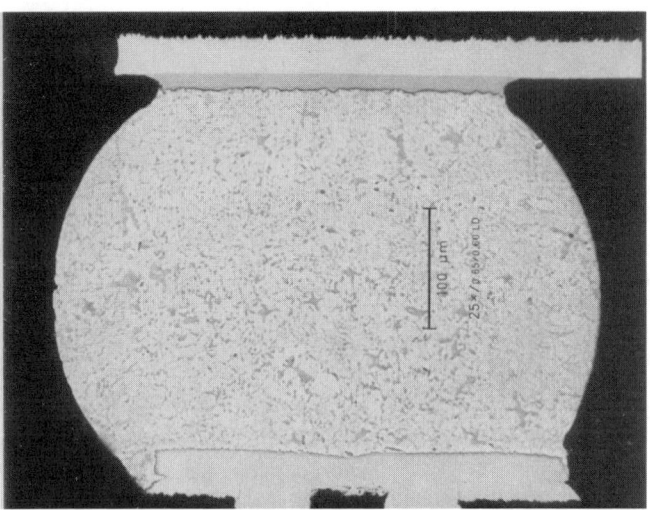

Fig. 2.1a. Characteristic shapes and microstructures of 169 CSP Solder Joints from NEMI Project on Pb-Free Assembly. (a) Sn-Pb alloy cross section

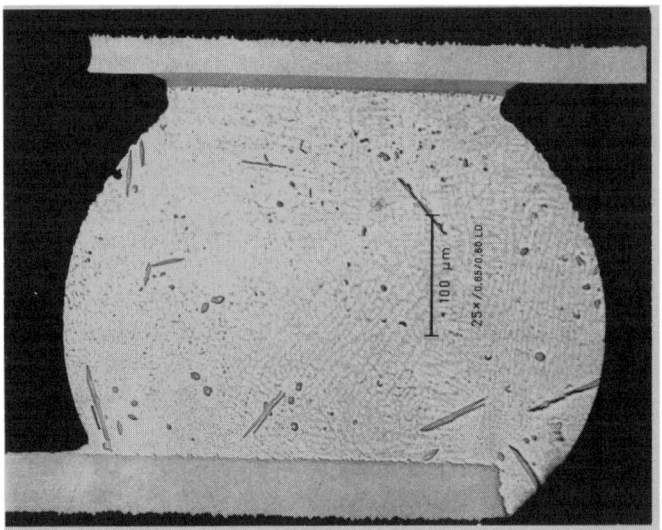

Fig. 2.1b. Characteristic shapes and microstructures of 169 CSP Solder Joints from NEMI Project on Pb-Free Assembly. (b) Sn-3.9Ag-0.6Cu alloy cross section

2.3 Solder Alloy Solidification and Microstructural Development

During their life cycles, from assembly through use in products, solder alloys undergo a wide range of phase transformations that change their compositions, their properties, and the compositions and properties of the materials they contact. Solder alloys melt, then wet and react with the board and component lead materials, sometimes causing significant dissolution of board and component lead materials. As the solder is cooled, the solid phases that form and their morphologies depend on the cooling rate and the relative ease of nucleation of solid phases in the melt and on pre-existing phases.

Solder alloys may exhibit metastable phase formation during solidification as well as microstructural coarsening and reactions with board substrates and component lead materials in the solid state. During product use, the solubilities and the distribution of phases change as a result of thermomechanical fatigue. In terms of the effect of solder alloy composition on the ability to assembly PWBs, some of these phase transformations are clearly identifiable and straightforward to analyze, particularly their melting behavior relative to Sn-Pb eutectic. Others require the full scope of materials science concepts and characterization tools to understand. The phase transformation and interface phenomena involved and typical characterization tools and methods used in solder alloy design are listed in Table 2.2.

2.4 Melting Behavior

2.4.1 Alloy Thermodynamics

The melting behavior of lead-free solder alloys has been, from the beginning, judged against the Sn-Pb system. The Sn-Pb phase diagram (Figure 2.2) is a simple binary eutectic phase diagram characterized by a liquid phase and two solid phases, each with substantial solid solubility. The eutectic composition is Sn-37Pb with a eutectic temperature of 183°C, a reduction of melting temperature (T_t) by almost 50°C from pure Sn at 232°C. The microstructure of Sn-37Pb on solidification is frequently used in metallurgy textbooks as an example of a classic eutectic structure, with its intermixed Sn and Pb solid solution phases.

Without exception, all Pb-free solders considered as candidates to replace Sn-Pb eutectic have been based on Sn modified with additional alloying elements. Using Sn as a base metal, alloying additions have been sought to reduce the liquidus temperature from 232°C while keeping the temperature

Table 2.2. Phase transformation and interface phenomena important for Pb-free solder alloy systems

Phase transformation or interface phenomenon	Property being evaluated	Characterization tool	Measurement signature of phase transformation
Melting, solidification, and phase transformation behavior	Thermodynamic properties of equilibrium phase transformations, including transformation temperatures, compositions	CALPHAD software tools, such as ThermoCalc[1], producing calculated phase diagrams	
Melting behavior	Temperatures where various phases begin to melt and where melting is complete	Differential scanning calorimetry (DSC), differential thermal analysis (DTA)	Endothermic peaks associated with melting of each phase
Solidification behavior	Temperatures where various phases begin to form from the melt during cooling	Differential scanning calorimetry (DSC), differential thermal analysis (DTA)	Exothermic peaks associated with phase formation
	Existence of metastable phase formation – stable phases do not nucleate easily	Differential scanning calorimetry (DSC), differential thermal analysis (DTA)	Exothermic peak well below equilibrium solidus temperatures followed by rapid rise in sample temperature up to equilibrium transformation temperature; i.e. recalescence

---

[1] In this chapter commercial products or trade names are identified for completeness, their use does not imply an endorsement by NIST.

		Differential scanning calorimetry (DSC), differential thermal analysis (DTA)	Exothermic peak well below equilibrium solidus temperatures indicating Scheil effect active during solidification
	Existence of metastable phase formation – slow diffusion in solidifying phase leads to composition gradients in solid and non-equilibrium amounts of remaining phases		
Microstructures of solid phases – including reaction interfaces between phases	Phases forming during trans-formations	Optical microscopy, transmission electron microscopy (TEM), scanning electron microscopy (SEM), electron probe microanalysis (EPMA)	Contrast differences in polished cross sections or as-formed structures; in SEM, TEM, and EPMA, composition maps; in TEM, diffraction characteristic of specific phases
		X-ray powder diffraction	Diffraction patterns characteristic of different phases
Wetting and solderability	Conditions for wetting of liquid solder on different substrates	Sessile drop experiments and wetting balance tests	Contact angle and force/time, respectively, as a function of alloy composition, substrate and its condition, flux, and other wetting conditions

difference between solidus and liquidus temperatures as small as possible, even using a eutectic alloy if possible. The thermodynamic characteristics of Sn with other alloying agents can be understood by considering the Sn-Bi, Sn-Ag, Sn-Sb and similar binary phase diagrams. (Binary alloy phase diagrams can also be found in the ASM Binary Alloy Phase Diagram Handbook. [24] Kattner et al. give an extensive discussion of the binary and ternary phase diagrams and sources for solder alloy phase diagrams. [25-26])

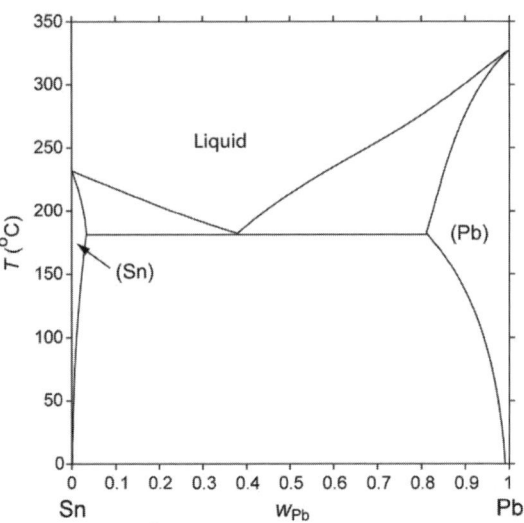

Fig. 2.2. Sn-Pb phase diagram.

In the simple Sn-Bi eutectic phase diagram (Figure 2.3), there is significant solid solubility of Bi in Sn, up to 22 wt% Bi in Sn at the eutectic temperature of 139°C. The liquidus temperature decreases with increasing Bi concentration from 232°C for pure Sn to 139°C at Sn-58Bi. The solidus temperature decreases with increasing Bi concentration from 232°C for pure Sn to 139°C at Sn-22Bi. Alloying elements that display significant solid solubility also exhibit significantly lower liquidus temperatures and eutectic temperatures than those with little solid solubility. Diagrams of this type include Sn-Zn (Figure 2.4) and Sn-Cd. This behavior is in contrast to the Sn-Ag diagram (Figure 2.5). There is negligible solid solubility of Ag in solid Sn. The liquidus temperature decreases from 232°C for pure Sn to 221°C at Sn-3.5Ag. The hypoeutectic solidus line is essentially a vertical line close to pure Sn. This behavior is characteristic of Sn-Cu (Figure 2.6), Sn-Ni (Figure 2.7), and Sn-Au (Figure 2.8).

The third characteristic phase diagram is illustrated by the Sn-Sb system (Figure 2.9). The Sn-Sb system contains a peritectic at the Sn-rich side of the phase diagram, leading to an increase in liquidus temperature with increasing Sb concentration. The phase field for tin, marked "Sn" on the left indicates significant solubility of Sb in Sn at temperatures above approximately 100°C. The Sn-In binary system (Figure 2.10) is slightly more complicated: in addition to the liquid and the two terminal solid solution phases, the Sn-In binary system contains two intermediate phases, both forming through peritectic reactions from the liquid and one of the terminal solution phases. At a temperature of 120°C, a liquid of Sn-51In decomposes by a eutectic reaction into the two intermediate phases.

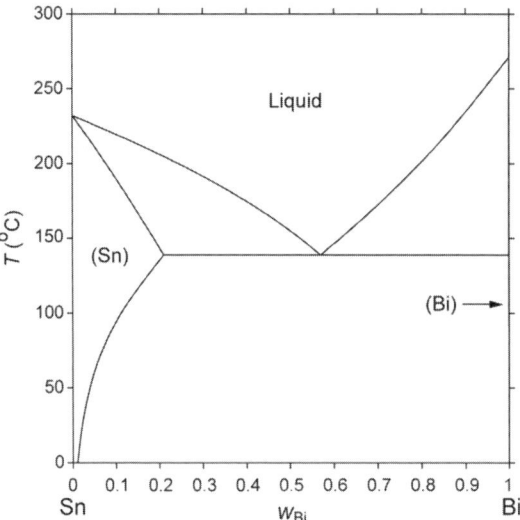

Fig. 2.3. Sn-Bi phase diagram.

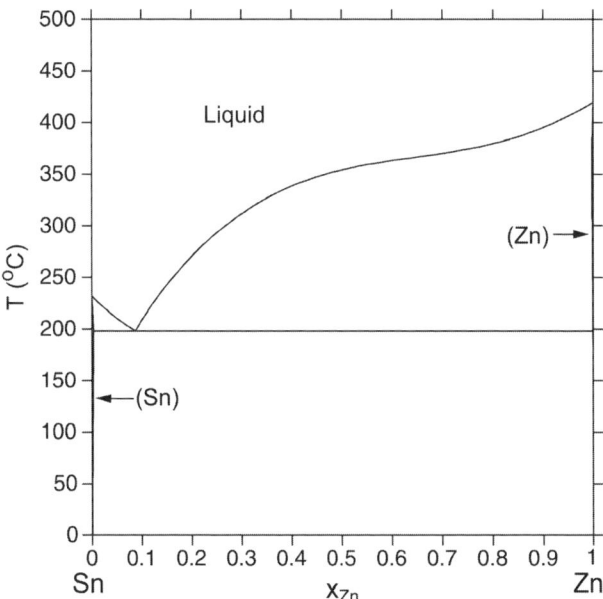

Fig. 2.4. Sn-Zn phase diagram.

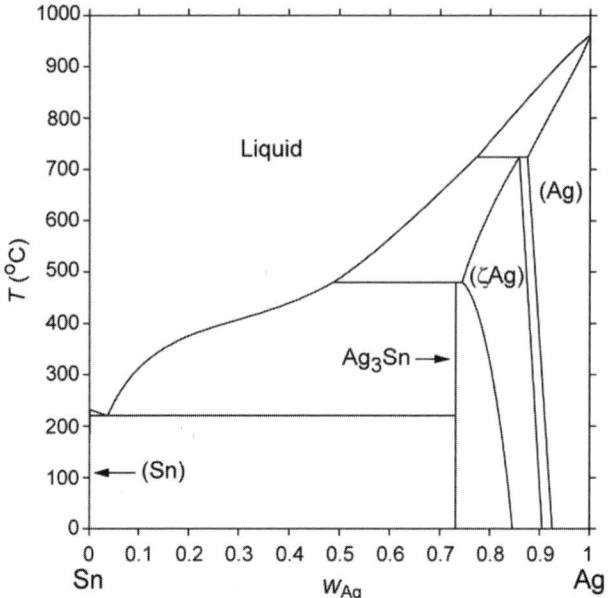

Fig. 2.5. Sn-Ag phase diagram.

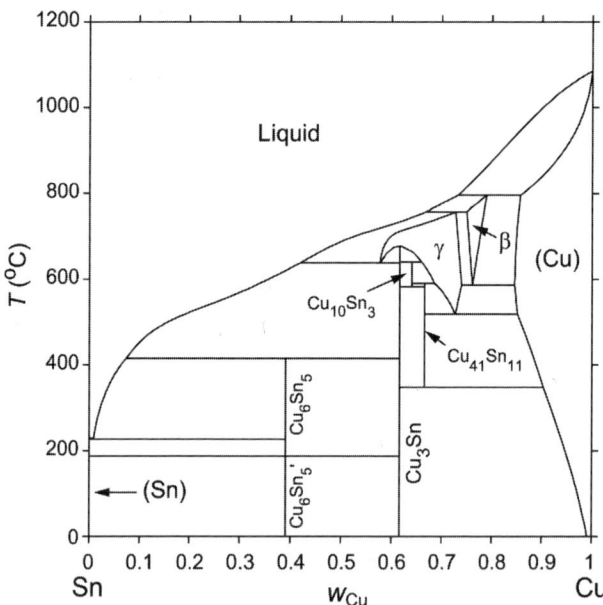

Fig. 2.6. Sn-Cu phase diagram.

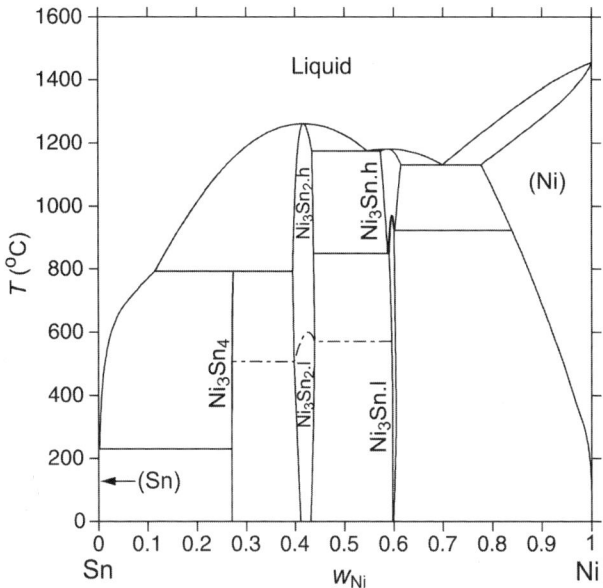

Fig. 2.7. Sn-Ni phase diagram.

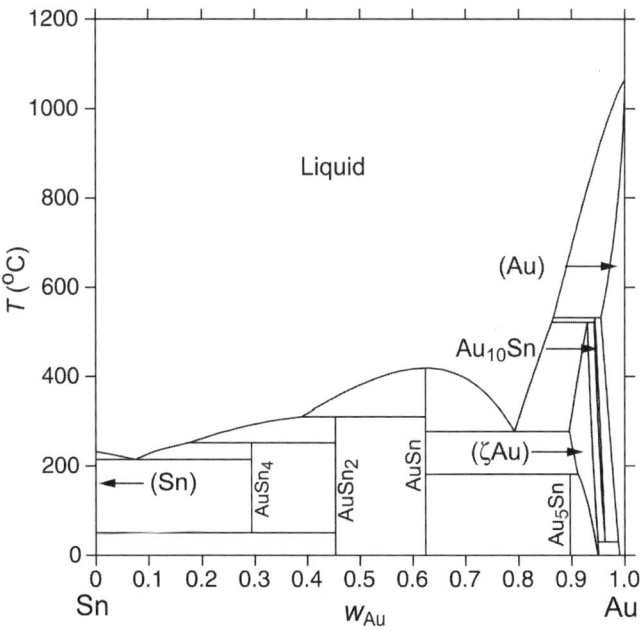

Fig. 2.8. Sn-Au phase diagram.

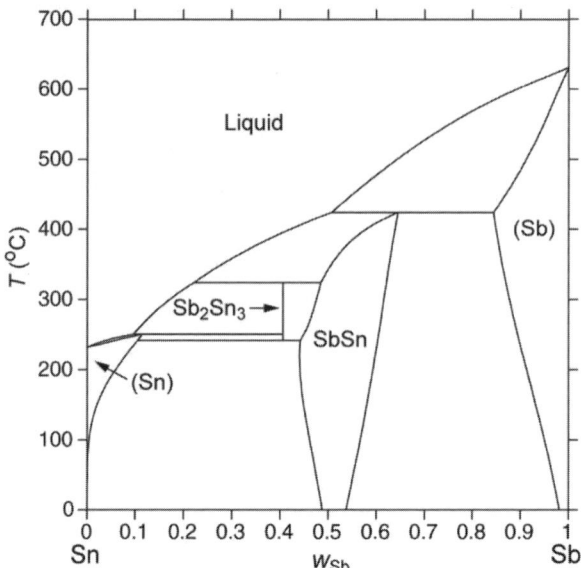

Fig. 2.9. Sn-Sb phase diagram.

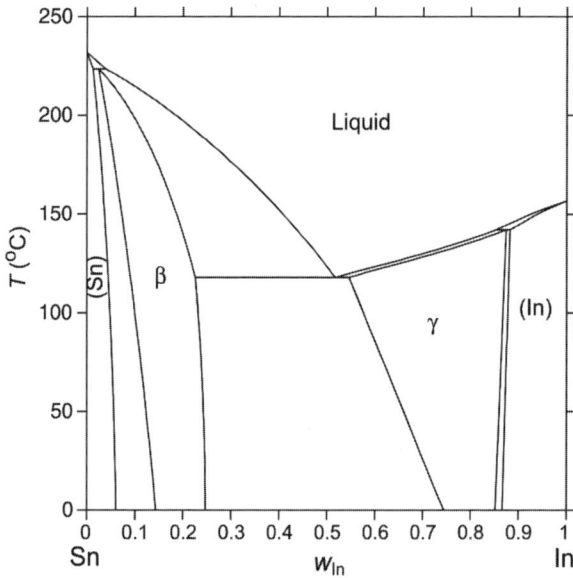

Fig. 2.10. Sn-In phase diagram.

These binary phase diagrams can be used to understand the melting behavior possible with ternary and quaternary Pb-free solder alloys. The primary alloy design criteria described above required (1) the liquidus temperature to be as close as possible to 183°C, in order to avoid changing manufacturing processes, materials, and infrastructure, (2) the solidus temperature to be as close as possible to the liquidus temperature, to avoid tombstoning phenomenon and fillet lifting (which are described further in this chapter), and (3) the solidus temperature significantly higher than the solder joint's maximum operating temperature. Binary and ternary eutectics obviously meet the second criterion; however, eutectic Sn-based alloys tend to fall into two temperature regimes with respect to the other two criteria. The high temperature, Sn-rich eutectics are Sn-0.9Cu (227°C), Sn-3.5Ag (221°C), Sn-0.16Ni (231°C), Sn-10Au (217°C), Sn-9Zn (199°C), and Sn-3.5Ag-0.9Cu (217°C) (Figure 2.11). The low temperature eutectic solders are Sn-58Bi (139°C), Sn-59Bi-1.2Ag (138°C) (Figure 2.12), Sn-59Bi-.04Cu (139°C) (Figure 2.13) and Sn-51In (120°C). (The eutectic in the Sn-Cd binary system of 177°C is close to ideal as a substitute for Sn-Pb from the point of view of melting point. However, Cd is highly toxic.)

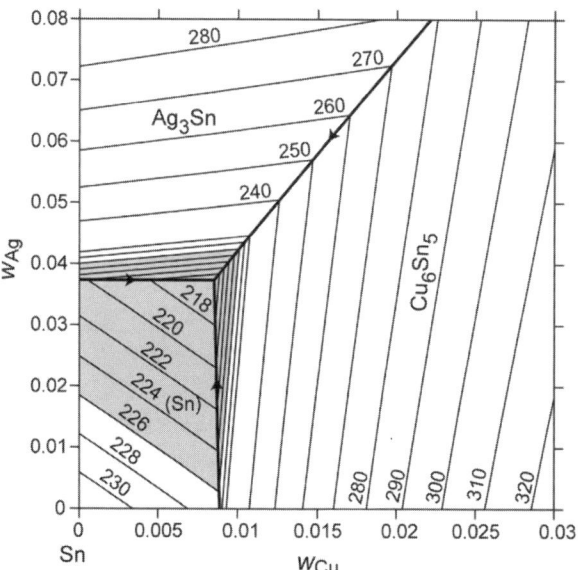

Fig. 2.11. Sn-Ag-Cu phase diagram – liquidus projection. [27]

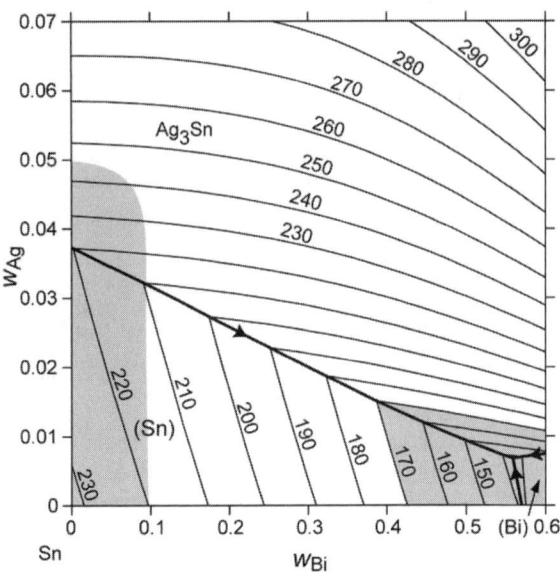

Fig. 2.12. Sn-Ag-Bi phase diagram – liquidus projection. [28]

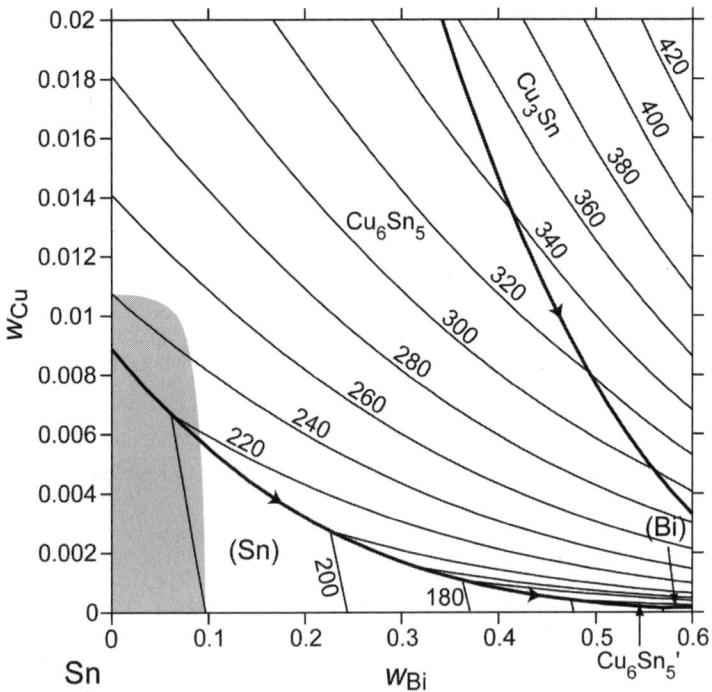

Fig. 2.13. Sn-Bi-Cu phase diagram – liquidus projection.

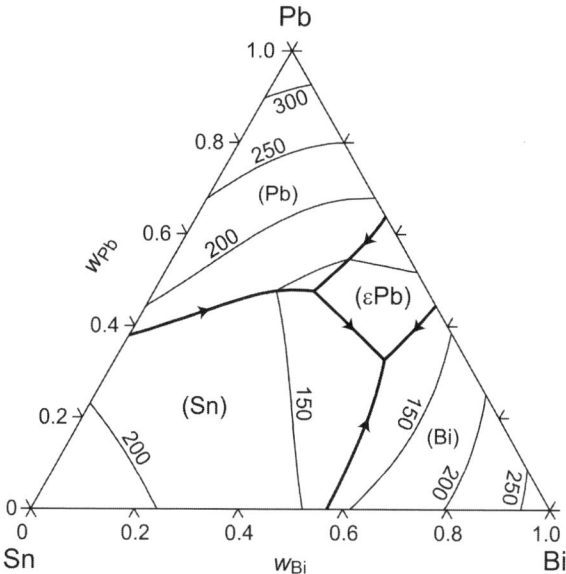

Fig. 2.14. Sn-Bi-Pb phase diagram – liquidus projection.

The NCMS Pb-Free Project member companies selected solders with liquidus temperatures less than 225°C and with an equilibrium pasty range (the difference between liquidus and solidus temperatures) less than 30°C. In the later IDEALS [11-14] and NEMI projects[15-18], candidate Pb-free solders were limited to eutectic and near eutectic, Sn-rich solders with smaller pasty ranges. Bismuth-containing solder alloys were eliminated as candidates for the primary replacement for the Sn-Pb eutectic alloy because of the possibility of forming the low temperature ternary eutectic at 96°C when a Bi-containing solder alloy is mixed with Sn-Pb surface finishes (Figure 2.14).

In the NCMS Project, it was shown that the composition dependence of the liquidus temperature for Sn-rich solder alloys could be estimated from a simple linear equation for additions of Ag, Bi, Cu, Ga, In, Pb, Sb, and Zn to Sn. [1] For Ag, Bi, Cu, and Pb, the coefficients were derived from the slopes of the Sn-X (X = Ag, Bi, Cu, Pb) binary phase diagram liquidus curves

$$T_\ell = 232°C - 3.1*W_{Ag} - 1.6*W_{Bi} - 7.9*W_{Cu} - 3.5*W_{Ga} - 1.9*W_{In} \qquad (2.1)$$
$$- 1.3*W_{Pb} + 2.7*W_{Sb} - 5.5*W_{Zn}$$

where the coefficients are in units of °C, and W_X is the amount of element X in wt.%. This equation is valid for the following alloy additions to Sn (expressed in wt.%): Ag < 3.5, Bi < 43, Cu < 0.7, Ga < 20, In < 25, Pb < 38, Sb < 6.7, and Zn < 6.

Using this equation, the maximum decrease from the melting point of pure Sn with additions of Ag and Cu was estimated to be 15-16°C, in agreement with the measured ternary eutectic temperature in the Sn-Ag-Cu system of 217°C, as seen in Figure 2.11. [27] From this simple equation, many alloy compositions with Bi, In, and Zn additions can be identified that exhibit liquidus temperatures of 183°C, the eutectic temperature of Sn-Pb eutectic solder. The problem with most of these alloys is that their solidus temperatures are significantly lower than 183°C. For example, the Sn-Bi binary alloy with a liquidus temperature of 183°C has a solidus temperature of 139°C. The issue of limiting the pasty range is particularly serious for through-hole wave joints, for which alloys with a large pasty range may exhibit fillet lifting as discussed further in this chapter.

One alloy that generated considerable interest was Sn-8Zn-3Bi with a liquidus temperature of 199°C and a pasty range of 10°C, thus having a significantly lower liquidus temperature than alloys in the Sn-Ag-Cu system. However, zinc-containing alloys have been observed to oxidize easily, showing severe drossing in wave solder pots, are prone to corrosion, and have a solder paste shelf life that is measured in terms of days or weeks compared to months for eutectic Sn-Pb. The bismuth is added to improve the wettability, reduce the liquidus temperature, and reduce corrosion compared with binary Sn-Zn alloys. The presence of bismuth may also result in the formation of low melting point phases in contact with Sn-Pb coated components and boards, affecting the reliability of the assembly as in the case of Sn-58Bi. Due to these issues, the use of this Sn-Zn-Bi alloy as a general replacement for eutectic Sn-Pb has been limited. [18]

The equilibrium melting behavior of four compositions in the Sn-Ag-Cu system that have been used as replacements for Sn-Pb eutectic solders is illustrated in Figures 2.15 and 2.16; with one additional high Ag/high Cu composition shown for comparison. These four solder compositions are Sn-3.0Ag-0.5Cu, Sn-3.5Ag-0.9Cu, Sn-3.9Ag-0.6Cu, and Sn-1.0Ag-0.5Cu. Figure 2.15 shows the total solid fraction as a function of temperature during heating for these four alloys. A comparison of these SAC alloys illustrates several important points regarding the sensitivity of the melting behavior to changes in composition. All of these alloys have a high fraction of Sn in the solid phase. For near-eutectic alloys, the solid alloy is >90% solid Sn that melts very close to 217°C. The total fraction of intermetallic phases over wide composition ranges is small as illustrated for the Sn-3.9Ag-0.6Cu alloy

(Figure 2.16a), and difficult to detect using standard DTA (differential thermal analysis) measurement systems.

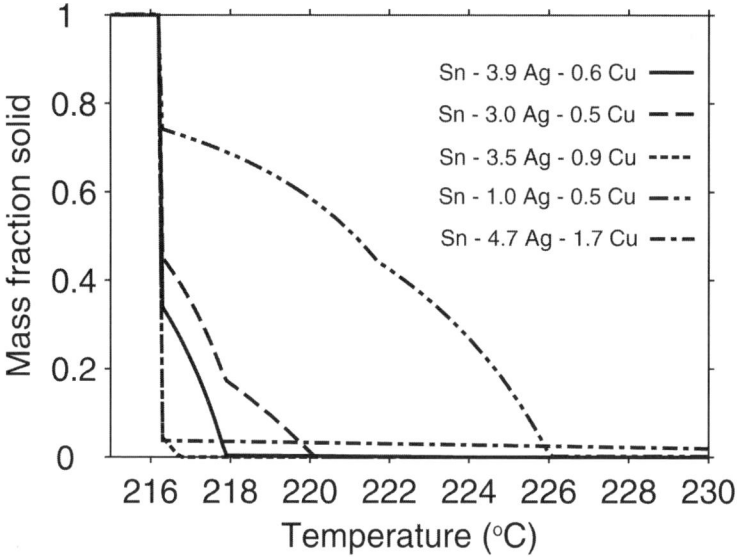

Fig. 2.15. Comparison of calculated fraction solid as a function of temperature for five different Sn-Ag-Cu alloys.

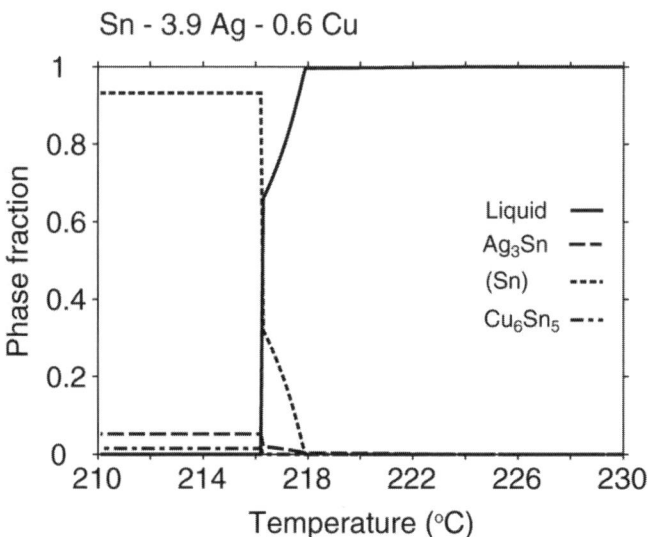

Fig. 2.16a. Calculated melting path for Sn-3.9Ag-0.6Cu.

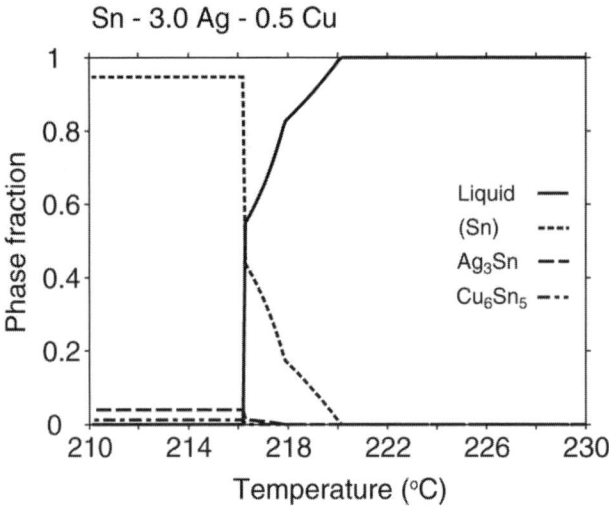

Fig. 2.16b. Calculated melting path for Sn-3.0Ag-0.5Cu.

The "effective" liquidus temperatures measured will, therefore, be 217°C for a wide range of compositions. In reflow soldering, it is likely that this small fraction of intermetallic phase will have a correspondingly small effect on solder flow and wetting, even if the solder in the joint never becomes completely liquid. For the two lower Ag alloys, the liquidus temperatures are 220°C (Sn-3.0Ag-0.5Cu) and 226°C (Sn-1.0Ag-0.5Cu). Since the fraction of Sn remaining between 217°C and the liquidus temperature is significant, the alloys must reach their liquidus temperature during assembly to properly reflow. This is illustrated in greater detail in Figure 2.16b with the equilibrium fractions of the individual phases, liquid, Sn, Ag₃Sn, and Cu₆Sn₅, shown as a function of temperature during heating for Sn-3.0Ag-0.5Cu.

Another useful representation of the melting behavior of SAC alloys as a function of temperature and composition is an isothermal section through the Sn-Ag-Cu phase diagram as presented in Figures 2.17 a-d. The experimentally determined eutectic composition of Sn-3.5(±0.2)Ag-0.9(±0.2)Cu is indicated by the black square in Figure 2.17a. When we consider the typical tolerance ranges of alloy compositions in solder pastes (±0.2), the possible melting range for the eutectic composition Sn-3.5Ag-0.9Cu becomes approximately 13°C. Likewise, when the typical tolerance of (± 0.2 wt.%) in the alloy composition is included, the NEMI alloy Sn-3.9Ag-0.6Cu, shown by the medium grey square has a possible melting range of 12°C. The third tin-silver-copper alloy Sn3.0-Ag-0.5Cu, shown by the light grey square, has a melting range of 5°C. The fourth tin-silver-copper alloy Sn-1.0Ag-0.5Cu, shown in the black framed white square, has a melting range of 11°C.

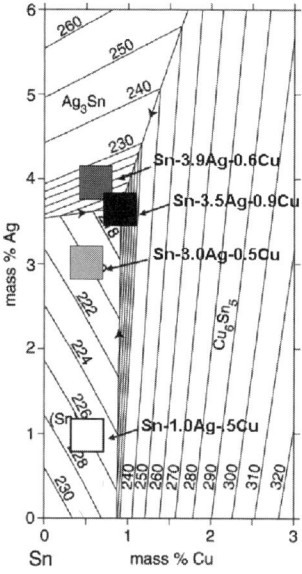

Fig. 2.17a. Liquidus projection of Sn-Ag-Cu diagram with four different alloys.

Figures 2.17b-d show the compositions over which there is <0.5% and <1% solid as the temperature is increased from 217°C, the eutectic temperature, to 219°C to 240°C to 270°C. In Figure 2.17b the region marked "L" and bounded by the black triangle is the range of compositions that are completely liquid at 219°C. The two regions outlined are compositions with less than 0.5% and 1% solid at temperatures higher than 219°C. The values of 0.5% and 1% were chosen since the presence of less than 1% solid is expected to have little or no effect on the reflow behavior of solder pastes. The remaining solid phase particles at this fraction are significantly smaller than the solder alloy powder particles from which they formed and will have a negligible effect on melting and coalescence of the alloy powders as they melt. As can be seen from Figure 2.17b, both the NEMI and the eutectic alloys have less than 1% solid remaining at 219°C. Beyond these two alloys, a wide range of alloys meet this criterion of having less than 1% solid remaining at 219°C. At 240°C (Figure 2.17c), the range of compositions with 0%, less than 0.5% and less than 1% solid remaining are extremely broad, encompassing all four alloys. [17]

The isothermal phase diagrams in Figures 2.17b-d can be used to estimate the change in solder composition of a Sn-Ag-Cu alloy held at 219°C, 240°C, or 270°C in contact with copper from the leads or pads, or silver from a board surface finish. For examples, Figure 2.17d shows the initial alloy composition of Sn-3.9Ag-0.6Cu and the final composition, Sn-3.85Ag-1.85Cu, as determined by the solubility limit of copper in the starting alloy at 270°C. The solder alloy Sn-1.0Ag-0.5Cu is shown for comparison; the composition of the final alloy saturated with copper at 270°C is Sn-0.99Ag-1.65Cu.

Based on the analyses described, the tin-silver-copper system is quite forgiving in terms of its insensitivity of melting behavior to composition over a wide composition range. Therefore, the effect of solder composition on reflow behavior should be minimal for compositions within this range. The same holds true for wave soldering. The temperatures for wave soldering are much higher than for reflow soldering and are determined by many factors, including the activity of the flux and the board design. The initial solder alloy composition will, however, affect how much copper and other metals will dissolve in the bath, so one might conclude that the base solder should contain high amounts of copper. A tradeoff in copper concentration actually occurs: For a given silver concentration, low initial copper concentrations encourage fast dissolution from the boards and the components, while high initial copper concentrations encourage intermetallic formation in colder sections of the bath. This tradeoff has led us to suggest a copper concentration limit in the as-received solder alloy of 0.5% to 0.6% for both wave soldering and reflow of electronic assemblies.

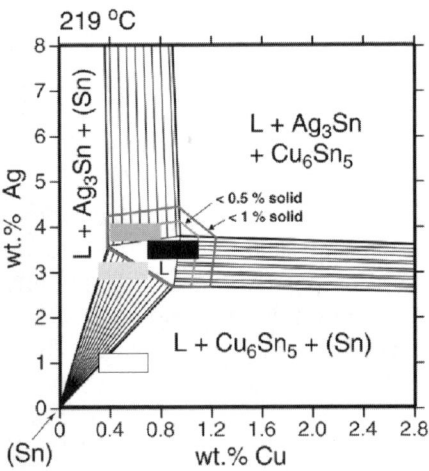

Fig. 2.17b. Isothermal section of Sn-Ag-Cu phase diagram at 219°C. Four compositions are shown with the ranges of composition allowable by industry standards corresponding to the borders of the rectangles. The region in composition space where <1% and <0.5% solid are outlined.

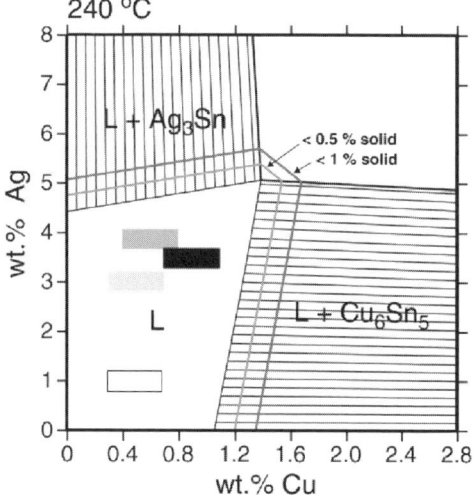

Fig. 2.17c. Isothermal section of Sn-Ag-Cu phase diagram at 240°C.

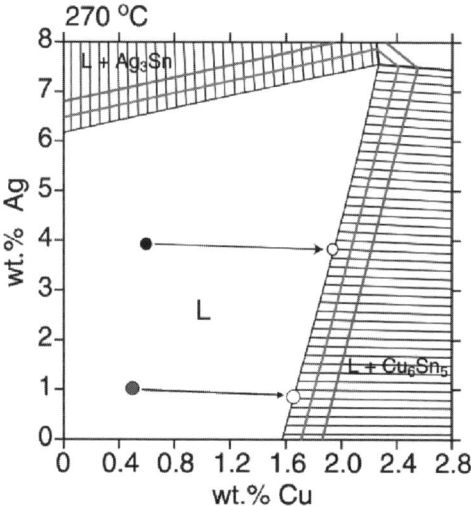

Fig. 2.17d. Isothermal section of Sn-Ag-Cu phase diagram at 270°C showing initial compositions Sn-1.0Ag-0.5Cu and Sn-3.9Ag-0.6Cu and their final compositions when they become exposed to Cu and their compositions become saturated with Cu at 270°C.

2.4.2 Copper and Nickel Dissolution into Molten Solder Alloys

The amount of copper and nickel from the substrate that dissolves into a molten solder and whether the alloy becomes saturated with respect to the substrate material are determined by the alloy thermodynamics and the dissolution kinetics. More insight into dissolution kinetics of copper in contact with SAC alloys can be found from the research of Chada et al. [30] who performed a comprehensive experimental study of the dissolution rate of Sn-3.5Ag as a function of composition and temperature for a controlled amount of solder in contact with copper. Figure 2.18 shows the effects of temperature, composition, time (t), and area of Cu-solder interface divided by the solder volume (A/V) on the dissolution of Cu in Sn 3.5Ag.

For the data presented here for the temperature range from 254°C to 302°C and for A/V values of 0.2 mm^{-1} and 0.4 mm^{-1}, the time for complete saturation of the Sn-3.5Ag molten alloy is short, on the order of 1 minute. This situation will exist for the limited solder volumes in surface mount solder joints and within the barrels of plated-through-hole wave joints. Therefore, depending on the time, the final molten alloy compositions will be between the initial alloy composition and their corresponding Cu-saturated compositions.

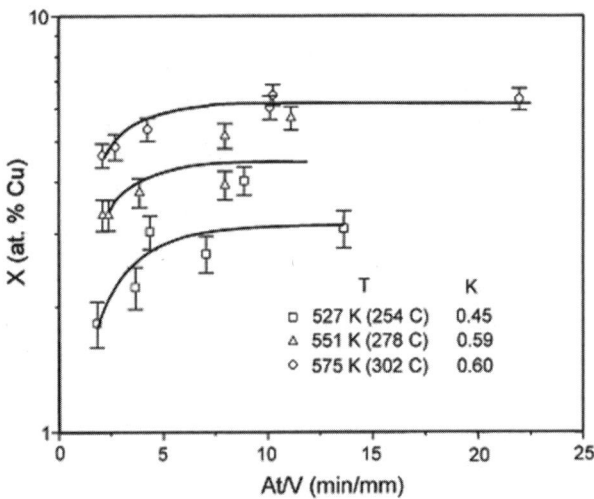

Fig. 2.18. Measurements of the average Cu content of isothermally reflowed specimens having two different substrate surface area (A) to solder volume (V) ratios and reflowed from 60 s, 120 s, 240 s. The curves represent fits of the Nernst-Brunner equation to the data. (From Chada et al. [30])

This situation is in contrast to essentially unlimited copper dissolution from the bottom-side of boards in contact with wave soldering or wave rework equipment, that is, the volume of the solder is essentially infinite compared with the area of contact between Cu and solder and the solder remains unsaturated with respect to copper.

Copper dissolution has been found to be a major factor limiting the time for wave rework of solder joints before the copper is completely dissolved. Dissolution of copper lands and barrels during rework of Pb-free plated-through hole solder joints using a wave solder process may lead to open circuits and electrical failure of printed wiring boards. Typical rework conditions for Sn-Pb eutectic solder allow for one to two rework operations using wave soldering rework equipment. The higher temperatures needed for wave rework of SAC, Sn-Ag, and Sn-Cu alloy through-hole joints lead to faster copper dissolution rates than with Sn-Pb eutectic solder. As discussed above, the Pb-free solder alloys are generally undersaturated with respect to copper and can therefore dissolve significantly more copper until they reach their solubility limits, as illustrated in Figure 2.17d for two Pb-free alloy compositions in contact with copper. Since the wave rework mini-solder fountain soldering bath used for rework has an essentially infinite volume compared to the amount of available copper that can dissolve into it, the bath will dissolve all available copper without saturation.

The presence of nickel in either a Pb-free SAC alloy or in the substrate seems to have a significant effect on the total amount of Cu, Cu-Ni or Ni substrate dissolving into the alloy and the rate at which it dissolves. Recent studies of the effect on substrate dissolution of Ni concentration in the alloy or the presence of Ni substrate instead of a Cu substrate agree with the earlier study of Korhonen et al. [31] Korhonen et al. determined that the dissolution rates of substrates with compositions between pure Cu and pure Ni into Sn-3.5Ag and Sn-3.8Ag-0.7Cu alloys depended strongly on the substrate alloy, as seen in Figures 2.19a and 2.19b. Similar results have been reported for dissolution of Cu substrates in the presence of a Ni-containing Pb-free alloy. It is thought that this effect has two primary causes: (1) the solubility limit of Ni in alloys in the Sn-Ag-Cu family is extremely low, much less than 0.1% leading to a fast local saturation of the molten alloy with alloys always close to Ni saturation and (2) the diffusion through and dissolution of the intermetallics that form on Ni, Ni-containing copper substrates, or on Cu substrates in the presence of Ni-containing solder are very slow. If the ability to rework plated-through-hole joints in wave soldering rework equipment is important, the addition of Ni to the alloy or the use of a Ni-containing substrate may provide additional flexibility in terms of rework time.

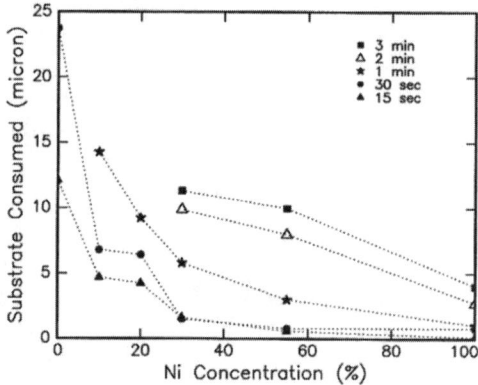

Fig. 2.19a. Substrate consumption vs. nickel content in the metal foil immersed in Sn-3.5Ag solder bath. [31]

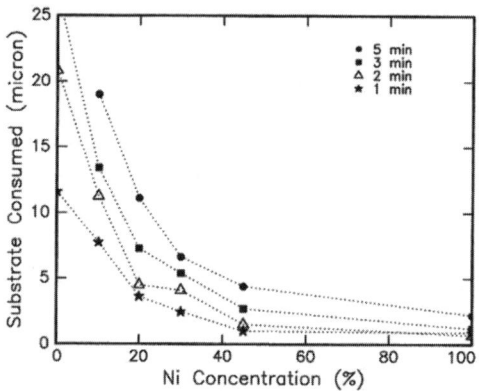

Fig. 2.19b. Substrate consumption vs. nickel content in the metal foil immersed in Sn-3.8Ag-0.7% Cu solder baths. [31]

2.5 Solidification Behavior

At first glance, it might seem reasonable to consider solidification during cooling from the liquid state as simply a reversal of the melting process, with the equilibrium solid phases forming at the equilibrium transformation temperatures. This is not the case for Sn-based alloys. There are five effects that we describe here that affect microstructural evolution: (1) difficulty of β-Sn nucleation, (2) dissolution of copper in the alloy changing the composition of the solidifying alloy; (3) precipitation of Cu-Sn intermetallics on the Cu-Sn interface during solidification; (4) non-equilibrium solids and liquids forming due to the Scheil effect, and (5) difficulty of liquid redistribution leading to fillet lifting. A recent review by Swenson [32] discussed in detail the various transformations and microstructural features that result from solidification of SAC alloys under a wide range of conditions. This section will be followed by a discussion of the effects of these phenomena on the performance and reliability of solder joints of various compositions, particularly the SAC alloys.

2.5.1 Effect of β-Tin Nucleation and Supercooling on Microstructural Evolution

It is well known that β-Sn has difficulty nucleating during cooling of solders from the process temperature. This difficulty in nucleation leads to formation of metastable phases, supercooling of the liquid phase well below where it is stable at equilibrium, and rapid formation of Sn dendrites once nucleation has occurred. These effects dominate microstructural evolution in SAC alloys since all of the SAC alloys being used as replacements for Sn-Pb are predominantly β-Sn, with small amounts of Cu_6Sn_5 and Ag_3Sn intermetallics.

During cooling from the completely molten state, the first solid phase forming in SAC alloys depends on the composition. For the composition Sn3.9Ag0.6Cu, the first phase that forms on cooling is Ag_3Sn The next phases expected from equilibrium thermodynamics calculations are Ag_3Sn and β-Sn. If β-Sn does not nucleate, the equilibrium phase diagram in Figure 2.11 cannot be used to predict the phases forming during cooling. A metastable phase diagram, Figure 2.20, must therefore be constructed to take into account the difficulty in nucleating β-Sn and the temperature at which β-Sn nucleation finally occurs. In the metastable transformation for Sn-3.9Ag-0.6Cu, Ag_3Sn continues to form as the supercooled liquid becomes depleted in Ag. The next stable phase to form in addition to Ag_3Sn is Cu_6Sn_5 when the composition reaches the metastable liquidus "valley" (line of two-fold saturation). The supercooled liquid continues to be depleted in Ag and Cu as the two intermetallics form. For alloys that start in the Ag_3Sn phase field, this means that there are large Ag_3Sn plates that form beginning at the liquidus surface and these continue to grow from the supercooled liquid below the eutectic temperature. At some temperature below the eutectic temperature, the supercooled liquid is sufficiently supersaturated to finally nucleate β-Sn in the presence of local heterogeneous nucleation sites in the solder. At this point, β-Sn forms as dendrites from a small number of nuclei and the solder

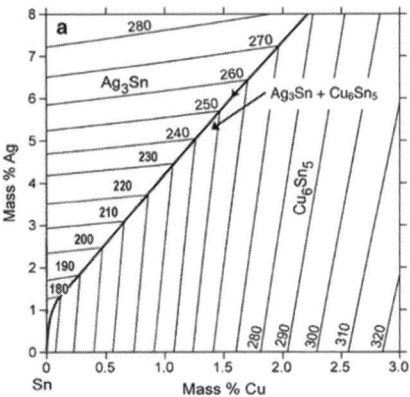

Fig. 2.20. Metastable Sn-Ag-Cu phase diagram showing no formation of β-Sn until a significantly lower temperature and with a lower Ag and copper concentration than the equilibrium ternary eutectic temperature and composition. (From Swenson, 2006 [32], recalculated using ThermoCalc)

joint solidifies completely with the remaining ternary alloy liquid forming regions with different phases and morphologies as discussed by Moon et al. [29]

The Sn-Ag-Cu phase diagram, the calculated solidification path, and DTA (Differential Thermal Analysis) results for the Sn-4.7Ag-1.7Cu are shown in Figures 2.11, 2.21a and 2.21b, respectively. At equilibrium, solidification begins with the formation of Cu_6Sn_5 at 265°C, at 238°C formation of Ag_3Sn begins and the remaining liquid should transform to a mixture of Sn, Ag_3Sn and Cu_6Sn_5 at the ternary eutectic of 217.5°C. However, in DTA experiments showing heat evolution during cooling (Figure 2.21b) from the liquid state, the first phases to form are Ag_3Sn and Cu_6Sn_5 at 244°C. Since β-Sn is difficult to nucleate, the liquid supercools by approximately 20°C while Ag_3Sn and Cu_6Sn_5 continue to form until the remaining liquid solidifies at 198.5°C. The latent heat (or heat of fusion) is released, leading to the solder self-heating to 217°C, the eutectic temperature. This phenomenon is known as recalescence. Supercooling can also be exhibited in the Sn-Pb system, but typically with a supercooling of 5°C.

This supercooling occurs even though large intermetallic particles are present which indicates that the intermetallics are ineffective as heterogeneous nucleation sites for β-Sn. As a result of the supercooling and the consequent rapid dendritic growth of β-Sn, the microstructure between the large intermetallic particles can be considered as practically an independent solidification process, forming the microstructure from an alloy with a composition much more Sn-rich than the original alloy. An analysis of the amount of supercooling (i.e., the temperature when β-Sn finally nucleates) and the slopes of the

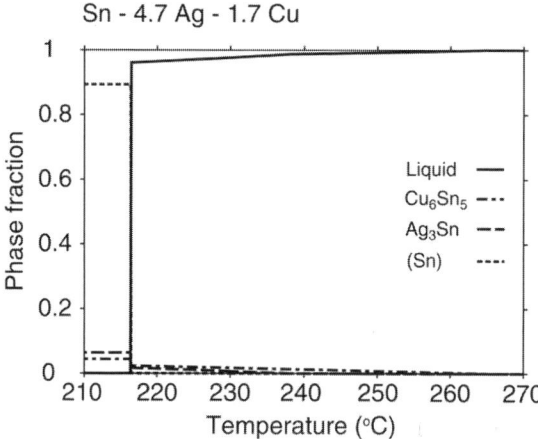

Fig. 2.21a. Calculated solidification path for Sn-4.7Ag-1.7Cu.

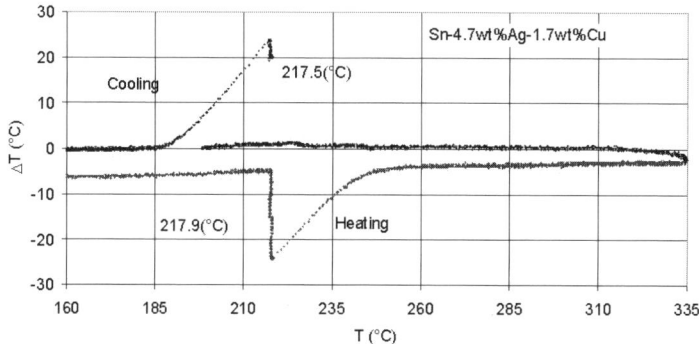

Fig. 2.21b. DTA heating and cooling curves for Sn-4.7Ag-1.7Cu. [27]

metastable liquidus surfaces allows the liquid composition at the start of growth of β-Sn dendrites to be estimated. Figure 2.22 show various micro-structures between β-Sn dendrites. Moon et al. determined that solidification sequences from a wide range of SAC alloys agree with that predicted for supercooling and solidification of a more Sn-rich composition existing at the time of Sn nucleation than that expected from equilibrium processes. [27]

The metallographic sections do not reveal these solidification paths in a straightforward manner. The primary Ag₃Sn phase can be recognized as elongated plates, usually seen edge-on. The plate morphology can be seen clearly in the etched structures from Jang and Frear [33] (Figure 2.23) and microstructures from the JCAA-JGPP study (34) (Figure 2.24). The Cu$_6$Sn$_5$ phase appears as more blocky particles in two dimensions which can be sections through rods or more equiaxed particles. The solidification sequences for the individual microstructural regions are not always apparent because of the initial difficulty in nucleating β-Sn and the subsequent rapid β-Sn dendrite growth once nucleation occurs.

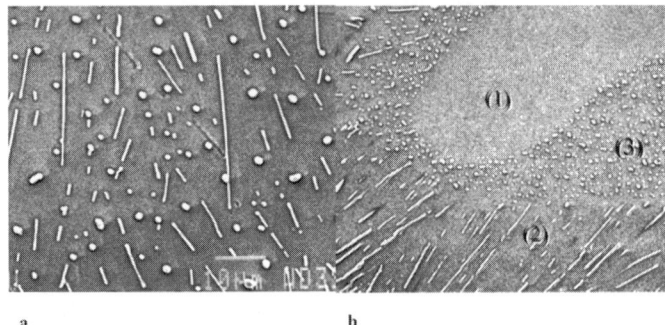

Fig. 2.22. SEM micrographs of eutectic structures: (A) ternary eutectic structure (matrix: (Sn), needle shape: Ag3Sn, and disk shape: Cu6Sn5). (B) region with co-existing (Sn)+Cu6Sn5 and (Sn)+Ag3Sn fine two phase regions near a (Sn) dendrite arm. Region labels 1 - (Sn); 2 - (Sn) + Ag3Sn; 3 – (Sn) + Cu6Sn5; 4 – (Sn) + Ag3Sn + Cu6Sn5. [27]

The best microstructural feature to separate the stages of solidification is the size of the largest intermetallic particles in various regions. An additional parameter that affects the phases that nucleate, their microstructures, and their resulting properties is the cooling rate, as demonstrated by Kim et al. [35] and Henderson et al. [36] Since cooling rate is not something that can be controlled in a practical way for complex assemblies, there are several other strategies that have been proposed to influence Sn-based alloys. Henderson et al. demonstrated that the formation of Ag$_3$Sn could be reduced and even suppressed by starting with a composition that is substantially lower in Ag than many common commercial alloys [36]. By using a typically observed supercooling of 20°C, Henderson et al. estimated that a SAC alloy must contain less than 2.7 wt% Ag to avoid formation of Ag$_3$Sn intermetallic plates. The intent to suppress Ag$_3$Sn formation before β-Sn forms has been the primary reason that the Sn-1.0Ag-0.5Cu SAC alloy has been proposed as an additional option to the higher silver compositions.

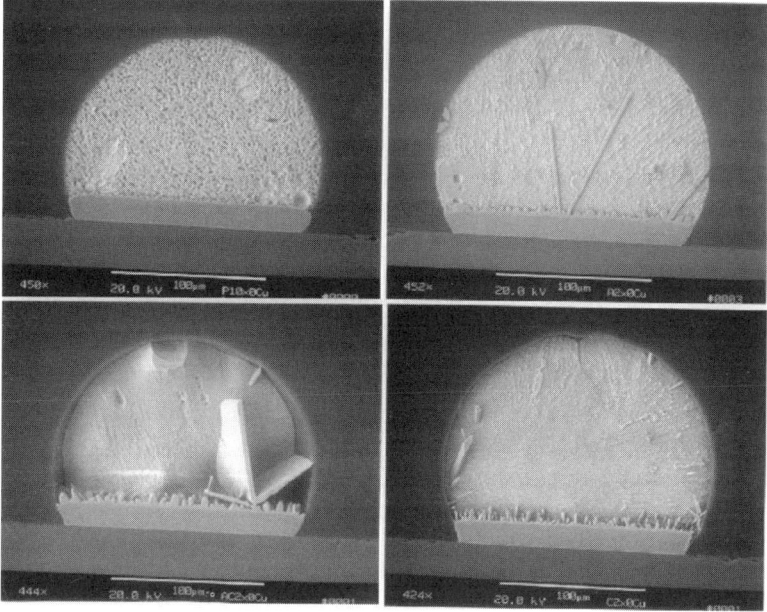

Fig. 2.23. Etched microstructures of Sn-3.5Ag showing Ag$_3$Sn platelets from unpublished research of Jang and Frear. [33]

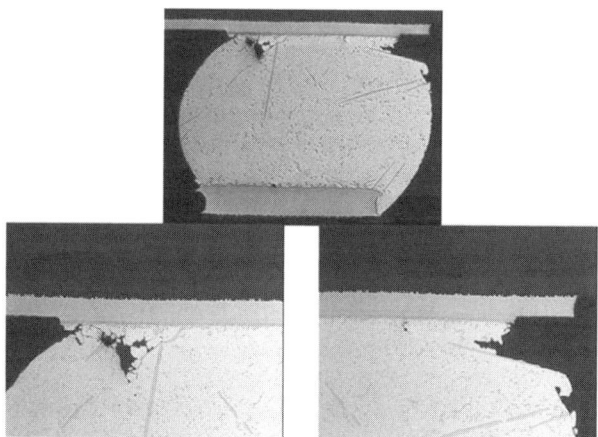

Fig. 2.24. Non-eutectic solidification behavior of Sn-Ag-Cu alloy can produce non-spherical solder ball shapes. These solder joints had very good reliabilities in thermal cycling under conditions from –55°C to +125°C. From the JCAA/JG-PP study. [34]

Another strategy to avoid Ag$_3$Sn plate formation and promote β-Sn nucleation has been to "inoculate" the SAC alloy with minor additions that promote nucleation of β-Sn at temperatures and compositions closer to their equilibrium values. [37-39] Swenson presents a critical analysis of the effects of Zn-additions on melting and solidification of Sn, Sn-Ag alloys, and SAC alloys. [32] More recent research by Anderson, Syed, and Dudek [40] suggest that other additives may also be effective at forming heterogeneous nucleation sites for β-Sn.

In any discussion of the role of Ag$_3$Sn plate formation on thermomechanical fatigue, it should be remembered that the presence of Ag$_3$Sn plates does not appear to have a direct influence on either failure path or solder joint lifetime [34, 36].

2.5.2 Role of Copper Substrate Dissolution on Alloy Solidification and Intermetallic Formation

An additional complication in analyzing the microstructure is the role of the board pad-solder and component pad-solder interfaces in both shifting the solder composition from the initial composition and being a preferred site for intermetallic formation. As noted in the previous discussion on melting, the solder composition can be shifted as a result of dissolution of substrate material in the molten solder. The amount and types of intermetallic that then form during solidification will be determined by the starting composition of the alloy and how much copper and other metals from the board and component have dissolved into the molten solder. For example, for the composition Sn-3.9Ag-0.6Cu at 270°C noted in Figure 2.17d, dissolution of copper in the solder joint can bring the overall composition to Sn-3.85Ag-1.85Cu when the liquid solder has dissolved the maximum amount of copper it can.

The formation of large Ag$_3$Sn plates depends on the composition in the joint being in the Ag$_3$Sn primary phase field. If, as a result of copper dissolution in the molten solder alloy, the copper concentration becomes high enough in the solder joint for the primary phase to be Cu$_6$Sn$_5$ rather than Ag$_3$Sn, then Cu$_6$Sn$_5$ will be the first phase that forms from the molten solder during cooling and not Ag$_3$Sn as in the original alloy composition. As the solder joint cools, Cu$_6$Sn$_5$ intermetallics form in the solder joint, both at the pre-existing intermetallics at the interfaces between the solder and the board and component metallizations and as rods in the solder itself.

2.5.3 Non-Equilibrium Effects of Diffusion in the Solid on Equilibrium Solidification

The pasty ranges based on equilibrium phase diagrams and the Lever Rule are the minimum pasty ranges that can occur during solidification. Segregation in the solid during solidification and metastable phase formation may extend these ranges in solder systems that exhibit substantial changes in the solubility of solid Sn during cooling. The amount of liquid present during cooling can be greater than predicted from the equilibrium phase diagram as a result of the well-known Scheil effect. In Lever Rule calculations, complete mixing in the liquid and complete diffusion in the solid at each temperature are assumed to have occurred during cooling to create the conditions of equilibrium. In Scheil calculations, complete mixing in the liquid is assumed but no diffusion in the solid is allowed during cooling. This leads to conditions only of local equilibrium at the solid-liquid interface during solidification, but not to equilibrium of the system.

Tin-based solder systems that exhibit this effect are Sn-based alloys containing bismuth, indium, and lead. For example, as a Sn-10Bi solder alloy is cooled from its liquidus temperature, the first solid that forms is Sn containing significantly less Bi than the Sn (Bi) solid solution at the eutectic temperature. If there is sufficient solid-state diffusion to maintain the equilibrium solid composition as the alloy cools, the final liquid transforms to a uniform solid solution at approximately 190°C, the equilibrium temperature and composition. If diffusion in the solid does not establish the equilibrium solid composition at each temperature as the alloy cools, the remaining liquid becomes increasingly Bi-rich, and has been observed to solidify at the eutectic temperature. For a Sn-6Bi solder, the liquidus temperature is approximately 224°C and the equilibrium pasty range is approximately 26°C; in the limit of no diffusion in the solid, the pasty range can be as large as 85°C. In the NCMS Pb-free Solder Project, DTA measurements of Sn-6Bi detected a measurable fraction of eutectic liquid that solidified at 139°C and, therefore, the non-equilibrium pasty range of 85°C as predicted by Scheil calculations [1-4].

The difference between the Scheil and Lever calculations is illustrated in Figures 2.25a and 2.25b for the ternary Sn-Ag-Bi system with calculations of the solid fraction as a function of temperature and composition. [Ref. 1-4] The liquidus projection of the ternary phase diagram is shown in Figure 2.12, where the lines correspond to compositions with the same liquidus

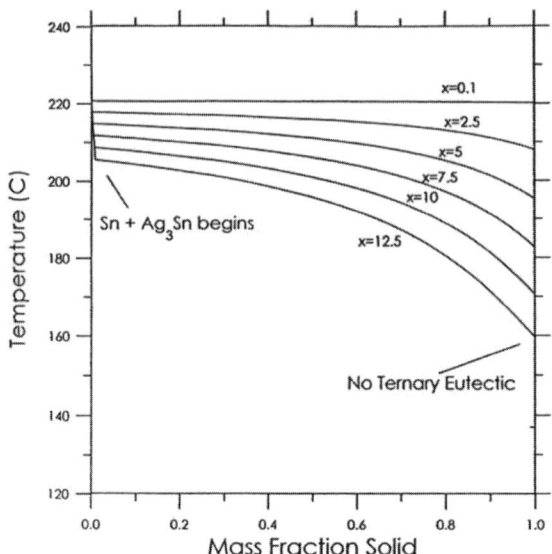

Fig. 2.25a. Lever solidification calculation for Sn-Ag-Bi. [1]

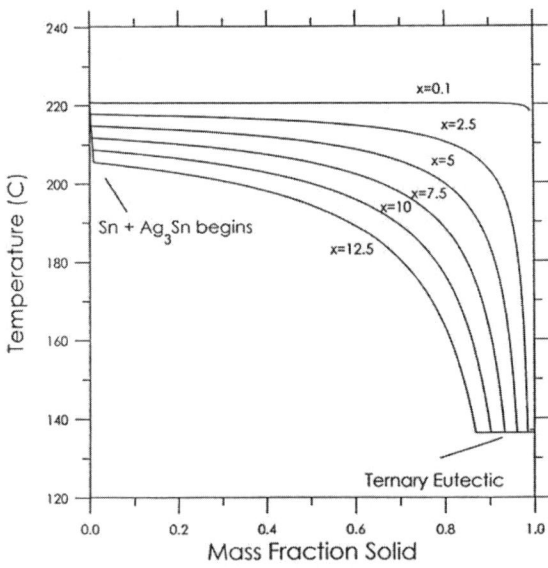

Fig. 2.25b. Scheil solidification calculation for Sn-Ag-Bi. [1]

temperatures. Considering the composition Sn-3.5Ag-7.5Bi, the last liquid solidifies at 185°C based on the equilibrium phase diagram; however, as a result of segregation during solidification, some liquid is predicted to still be present until the ternary eutectic temperature of 138°C. The amount of non-equilibrium liquid present depends on the cooling conditions and will be between the limits defined by the two curves for Sn-3.5Ag-7.5Bi in Figures 2.25a and 2.25b. In systems like Sn-Ag-Cu, there is little solubility of Ag and Cu in Sn and non-equilibrium solidification due to interdiffusion in the solid and the Scheil effect play a minor role in the behavior of SAC alloys.

2.5.4 Fillet Lifting

Fillet lifting is a defect-formation phenomenon for wave soldered through-hole joints that occurs during solidification for some Pb-free solders that does not typically occur for eutectic Sn-Pb. Fillet lifting, as shown in the micrograph in Figure 2.26, is characterized by the complete or partial separation of a solder joint fillet from the intermetallic compound on the land near the shoulder of the through hole. This phenomenon was first identified in 1993 by Vincent and co-workers [8-10] in the UK DTI (Department of Trade and Industry)-sponsored Pb-free solder project. [8] In that study fillet lifting was attributed to the presence of the Sn-Bi-Pb ternary eutectic (98°C) resulting

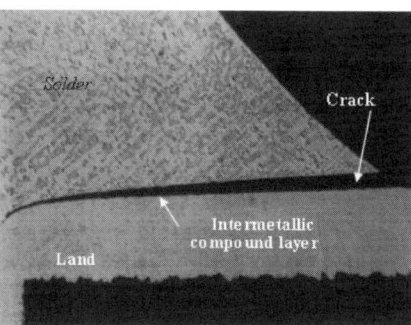

Fig. 2.26. Optical microscope cross section of fillet lifting in a through hole solder joint with Sn-3.5Ag-5Bi alloy. [1]

from Pb contamination of Bi-containing solders from the Sn-Pb hot-air-solder-leveled (HASL) board finish. While this can create the conditions for fillet lifting, this effect is now known to occur without Pb contamination for many Pb-free solder alloys, particularly for alloys containing Bi.

Suganuma, Boettinger et al. and Takao and Hasegawa have determined that fillet lifting is the result of "hot tearing," a mechanism that leads to the relief of thermally induced stresses *when the solder is between 90% and 100% solid.* [41-43] The differential shrinkage due to CTE mismatch between the board and the solder generates stresses in the circuit board and the solder as the solder solidifies. At lower solid fractions during solidification, fluid flow occurs relieving the stresses. As the volume fraction of liquid decreases, the stresses are carried by the dendritic matrix and failures occur at the weakest point, typically the location with the highest remaining liquid fraction: the board-side intermetallic compound/solder interface. The tendency for fillet lifting increases as the pasty range increases and the temperature difference between 90% and 100% solid ($\Delta T_{90\% \to 100\%}$) increases. It is typically worse for alloys with a large non-equilibrium pasty range, like Sn-Bi or Sn-Ag-Bi. Takao and Hasegawa[43] have quantified the tendency to fillet lifting as a function of alloy composition in terms of the enthalpy (energy) change as a function of composition and temperature during cooling which corresponds directly to $\Delta T_{90\% \to 100\%}$.

In the NCMS Pb-free solder project, "hot tearing" as the origin of fillet lifting was tested by taking Sn-3.5Ag, an alloy that showed minimal fillet lifting, and transforming it into an alloy showing close to 100% cracked joints with the addition of 2.5 wt% Pb. [1] The addition of 2.5% Pb increased the pasty range from 0°C to 34°C. The NCMS study predicted that Pb contamination from Sn-Pb surface finishes would lead to fillet lifting in alloys that in their uncontaminated state showed little or no fillet lifting. Subsequent large-scale industrial wave soldering trials revealed fillet lifting in through-hole joints with Sn-Ag, Sn-Cu, or Sn-Ag-Cu solders when combined with Sn-Pb surface finished components and/or boards. [44-45] It should also be remembered that Sn-Ag, Sn-Cu, and Sn-Ag-Cu fillets may also show fillet lifting for very thick boards without Pb contamination, as indicated in the NCMS Pb-Free Solder Project. [1] Fillet lifting under these conditions is a result of the supercooling, dendritic growth, and recalescence behavior described previously. Fillet lifting appears to be a cosmetic effect only: there have been no reported solder joint failures attributed to fillet lifting.

2.5.5 Solidification and Surface Porosity

As discussed above, solidification of high-Sn Pb-free alloys occurs with the formation of Sn dendrites as seen in the as-solidified structure in Figure 2.27a. The formation of Sn dendrites is accompanied by the redistribution of the interdendritic liquid and, ultimately, by a retraction of the remaining liquid as it redistributes and solidifies. The volume of the liquid is larger than the volume of the solid; solidification, therefore, leads to shrinkage and formation of porosity. This retraction of the interdendritic liquid leads to a rough outer surface in Figure 2.27a and, correspondingly, to an overall greater surface roughness than Pb-Sn eutectic, as seen in the SEM micrograph in Figure 2.27b. These micrographs indicate why the surfaces of properly soldered Pb-free solder joints appear significantly rougher than equally well-soldered Pb-Sn solder joints and with the visual inspection criteria for Pb-free solder joints discussed in the industry standard document IPC-A-610D (Acceptability of Electronic Assemblies) to indicate that a duller, rougher appearance for lead-free solder joints is not grounds for rejection. The dendritic microstructure causing the roughness is an intrinsic characteristic of the Pb-free alloys. The scale of the surface roughness depends

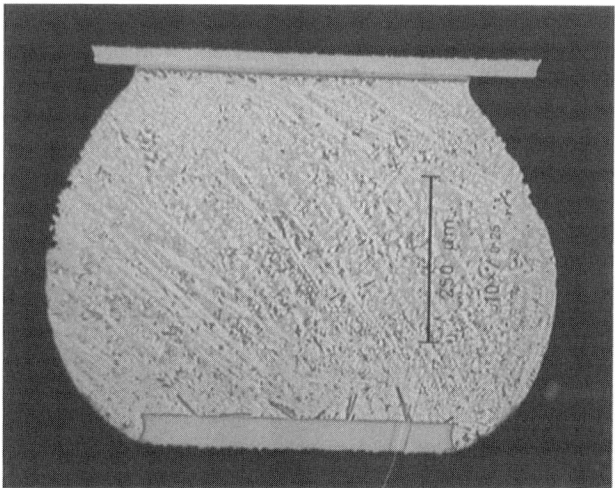

Fig. 2.27a. Cross section showing surface roughness of Sn-Ag-Cu alloy as solidified. [18]

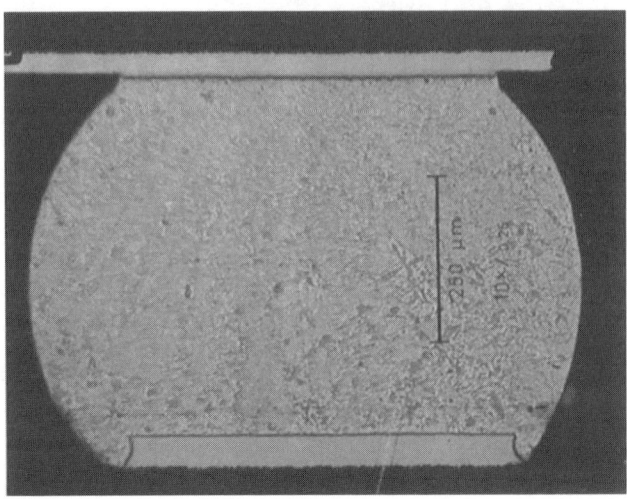

Fig. 2.27b. Cross section showing smooth surfaces of Sn-Pb alloy as solidified. [18]

on several factors, including the final joint composition and the cooling rate of the joint. The IDEALS project examined the effect of these surface "pores" caused by retraction of the solidifying interdendritic liquid on where the solder joint fails during thermal cycling. They determined that the surface "pores" between the dendrites were not preferential sites for solder joint failure. [14] Failure analysis of solder joints from a wide variety of components on thermally cycled assemblies in the NEMI Lead-Free Assembly Project did not show an effect of surface porosity on failure path or crack nucleation. [18] Hillman and his colleagues in the JCAA/JG-PP No-Lead Solder Project Report on thermal cycle testing also reported that there was no effect of porosity or position of Ag_3Sn plates on thermal cycle reliability, even for the extreme microstructures shown in Figure 2.24. [34]

2.5.6 Contamination of Pb-Free Solders

The use of a solder-coated board or component surface finish with a different composition than the solder paste or wave soldering alloy may result in different properties of the solder joints than expected from solder joints made from the paste or wave composition alloy alone. For Pb-contamination in particular, the liquidus temperature decreases by 1.3°C (per mass fraction Pb*100), which can be calculated using Equation 2.1. The next question to be answered is how the Pb contamination affects the solidus temperature, the lowest temperature where liquid exists. When Pb-free solder alloys are contaminated by Pb from the pre-tinned layer, the last liquid that solidifies may form a low melting eutectic. This case was studied in detail for Pb-contaminated Sn-Bi solders by Moon et al. using DTA methods in conjunction with calculations of the equilibrium phase diagram and Scheil solidification. [29] They found that contamination of Sn-Bi eutectic, Sn-5Bi, and Sn-10Bi alloys by 6% Pb results in the resulting alloy following the Scheil solidification path for liquid and solid and, hence, a measurable fraction of low melting Sn-Bi-Pb eutectic forms in these alloys at 95°C because the resulting alloys follow the Scheil path during solidification. In contrast, a similar study on the effect of Bi contamination on solidification of Sn-Pb solder alloys found introduction of small amounts of Sn-Bi from component surface finishes into Sn-Pb eutectic solder paste result in alloys following more closely the Lever Rule path. [46] The solidus and liquidus temperatures are slightly lower than those of the uncontaminated solders but there is no discernable amount of low melting point Sn-Pb-Bi ternary eutectic.

Since the freezing ranges of other Pb-free solder alloys may be similarly susceptible to Pb contamination, the freezing behavior of four additional solder alloys was studied by Kattner and Handwerker using Lever rule and Scheil freezing path calculations of the original solder alloy and the contaminated solder. [26] The level of contamination was chosen to be 6% Pb from the Moon et al. estimate of a Pb-concentration of 6% (mass fraction) in the solder from contamination by the component lead and board pre-tinning. [29] The original solder compositions and those resulting from Pb-contamination are listed in Table 2.3.

Table 2.3. Modified solder compositions as a result from contamination with 16% of 63% Sn-37% Pb solder. (Compositions are in % of mass fraction)

Original Solder Composition	Contaminated Solder Composition
Sn - 3.5% Ag	Sn - 2.9% Ag - 6% Pb
Sn - 4% Ag - 1% Cu	Sn - 3.4% Ag - 0.8% Cu - 6% Pb
Sn - 3.5% Ag - 4.8% Bi	Sn - 2.9% Ag - 4% Bi - 6% Pb
Sn - 3.4% Ag - 1% Cu - 3.3 % Bi	Sn - 2.8% Ag - 0.8% Cu - 2.8% Bi - 6% Pb

The calculations were carried out using the NIST solder database [47], the Thermo-Calc software package [48] and the Scheil and Lever programs [49]. Figures 2.28a-b show the calculated fraction solid as a function of temperature for Sn-3.5Ag and Sn-4Ag-1Cu and for the corresponding alloys contaminated by 6% Pb. Contamination of the binary eutectic alloy Sn-3.5Ag with 6% Pb lowers the liquidus temperature from 221°C to 213°C and lowers the solidus from 221°C to 177°C, creating an alloy with an equilibrium pasty range of 44°C. Likewise, Pb contamination of Sn-4Ag-1Cu alloy leads to a 2°C increase in the liquidus temperature from 225°C to 227°C and a decrease in solidus temperature from 215°C to 177°C.

2.6 Wetting and Solderability

Wetting of a liquid on a solid is determined by the relative energies of the liquid-vapor surface tension, the solid-liquid interfacial energy, and the solid-vapor interfacial energy, as represented by Young's Equation. The thermodynamics of an alloy play a central role in determining its intrinsic liquid-vapor surface tension. It is well known that the surface tension of pure Sn is significantly higher than Sn-Pb eutectic, as measured by White

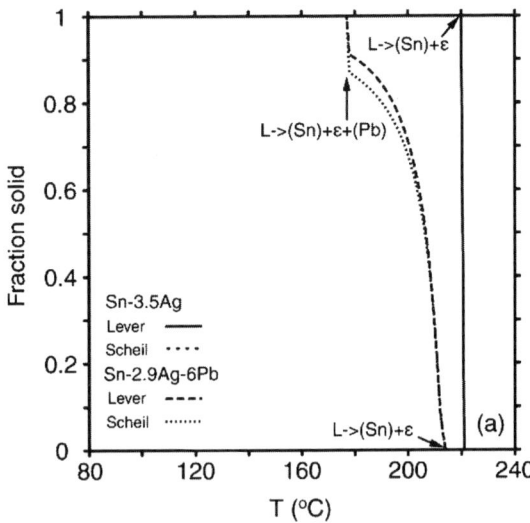

Fig. 2.28a. Lever and Scheil calculations for fraction solid as a function of temperature for Pb-free solders without and with 6% Pb contamination for Sn-3.5Ag. [26]

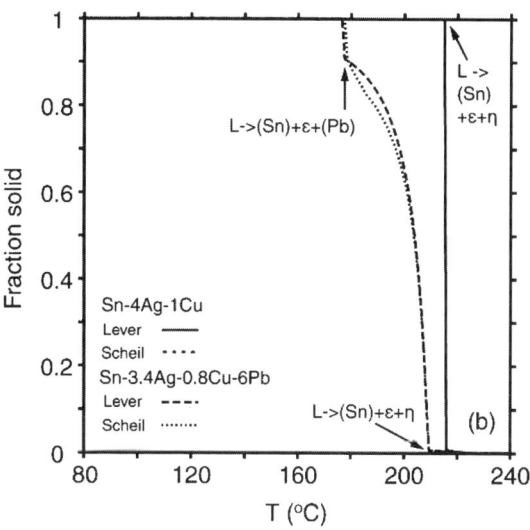

Fig. 2.28b. Lever and Scheil calculations for fraction solid as a function of temperature for Pb-free solders without and with 6% Pb contamination for Sn-4Ag-1Cu. [26]

as a function of temperature from pure Sn to pure Pb. [50] Ohnuma and his colleagues have used thermodynamic parameters to predict the surface tension and viscosity of the Sn-based liquids as a function of composition. [51-52] The higher surface tension of Pb-free high tin alloys over Sn-Pb alloys translates into generally higher contact angles for Pb-free alloys, less bridging between component leads, and a smaller process window for Pb-free solders than for Sn-Pb eutectic.

Evaluation of solderability in manufacturing has considerably greater complexity than wetting of an unreactive liquid on a substrate in a controlled laboratory environment as assumed by Young's equation [53]. Simple wetting balance and area-of-spread measurements are useful for separating the effects of solderability factors, such as flux, temperature, and substrate conditions, and for screening materials and soldering conditions when comparing different solder alloys. Manufacturing issues are discussed in detail in Reflow, Wave and Rework Chapters of this book Through numerous national and international Pb-free solder R&D projects using wetting balance measurements, solderability was found to be a major issue in discriminating between solder alloys only for Zn-containing alloys, and then only for concentrations greater than 1% Zn. [1]

For Pb-free alloys not containing Zn, their wetting characteristics on a specific metal substrate depend on the composition of the solder and the substrate, the temperature of the solder and the substrate, the size and thermal conductivity of the substrate, the liquidus temperature of the solder, the surface condition of the substrate, the gaseous experimental environment (oxygen, air, nitrogen), and, last but not least, the flux. A comparison of wetting balance data for various Pb-free solder alloys on copper from the IDEALS and NCMS projects indicates that: (1) in general, the temperature for similar wetting balance performance to eutectic Sn-Pb scales with the liquidus temperature of the Pb-free solder, and (2) the effect of the variables listed above are separable. Figure 2.29 from the IDEALS Pb-Free Project shows that the time to 2/3 wetting force for five Pb-free solder alloys compared with Sn-40Pb at three temperatures per alloy, $T_\ell + 25°C$, $T_\ell + 35°C$, and $T_\ell + 50°C$. [11-14] With the exception of Sn-0.7Cu, the character-ACTIEC 5 flux with 0.5% Cl which aids wetting (Figure 2.29a). When the flux is changed to pure Rosin flux with 0% Cl, four of the five Pb-free solders are again virtually identical to Sn-40Pb (Figure 2.29b). Only Sn-0.7 Cu-0.5Bi shows significantly pooree wetting than the other five solders.

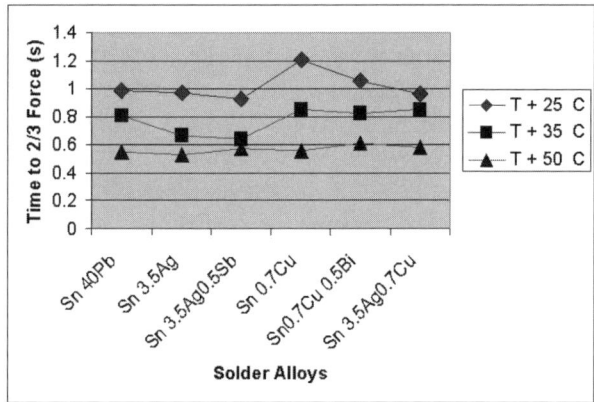

Fig. 2.29a. IDEALS project wetting data: wetting balance parameter, time to 2/3 force as a function of alloy composition for three different temperatures relative to the liquidus temperature for Actiec 5 flux. [11-14]

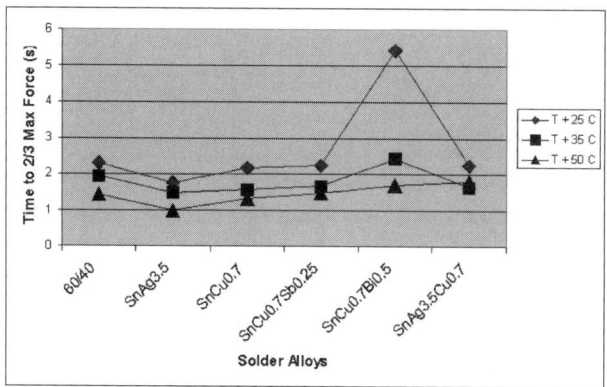

Fig. 2.29b. IDEALS project wetting data: wetting balance parameter, time to 2/3 force as a function of alloy composition for three different temperatures relative to the liquidus temperature for rosin flux. [11-14]

Bradley and Hranisavljevic have provided a thermodynamic basis for understanding the effect of alloy composition and surface finish/lead metal composition on wetting behavior based on how the liquidus temperatures change with dissolution of the surface finish or lead material into the solder. [54] They determined the temperatures at the start of solder paste coalescence and at full wetting, when all the solder particles are completely coalesced into a molten solder mass, for the alloys Sn-3.5Ag, Sn-3.8Ag-0.7Cu, Sn-1Ag-3Bi, Sn-1Ag-4.8Bi, and Sn-1Ag-7.5Bi and for four surface finishes. Figure 2.30

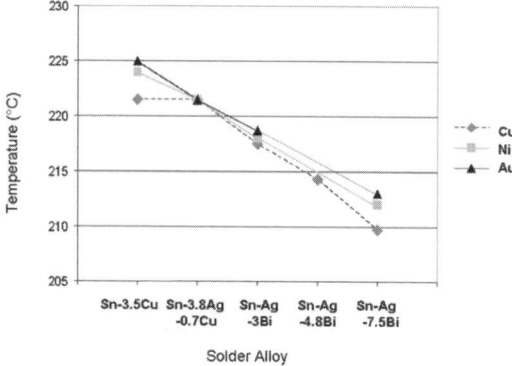

Fig. 2.30. Temperature at which solder paste begins to coalesce as a function of PCB substrate finish. Note the effect of Cu on depressing the coalescence temperature and therefore the effective wetting temperature of alloys without Cu. [54]

shows that the measured coalescence temperatures for all the solder alloy pastes, except for Sn-3.8Ag-0.7Cu, decrease with dissolution of the surface finish metal into the solder, with copper surface finish producing the greatest effect, followed by gold. This decrease in solder coalescence temperatures corresponds to a decrease in the liquidus temperatures produced by dissolution of the surface finish into the solder alloy. The alloy Sn-3.8Ag-0.7Cu is close to the ternary eutectic composition and the solidus temperature for this alloy is therefore unaffected by Cu dissolution.

2.7 Pb-Free Solder Mechanical Behavior and Solder Joint Reliability

2.7.1 Microstructure and Compositi on Effects on Solder Joint Mechanical Behavior and Reliability

The reliability of solder joints in assemblies is considered in more detail in the Reliability Chapter of this book. The intention of this section is to provide some physical insight into the composition and microstructural basis for the differences in mechanical and accelerated thermal cycling behavior of Pb-free alloys and Sn-Pb eutectic solder. A typical approach to estimating solder alloy and solder joint reliability is to rank alloys based on mechanical property values obtained in a particular test for a specific test condition or for a particular application. This testing can take the form of accelerated thermal cycling of a component solder joint or a simulated solder joint, creep at a particular temperature for different applied stresses or imposed strain rates, or even room temperature tensile stresses. The problem that has emerged is that the ranking of different Pb-free alloys with respect to Sn-Pb and to each other can change dramatically depending on the test condition chosen to be "representative."

The essence of the problem has been described well and in detail by J. P. Clech for Sn-Pb as well as Pb-free solders in the following statement [55]:

"The mechanical behavior of solder depends on the joint microstructure and is affected by many parameters such as intermetallics, joint or specimen size, cooling rate of the assembly after soldering, aging in service, etc. Test factors such as specimen or load eccentricity, temperature variations and measurement errors also contribute to the scatter in the mechanical properties of solder as is well known, for example, for steady state creep."

Alloy solder joints are composed of heterogeneous microstructures with features that vary from the tens-of-micrometers scale to the sub-100 nm nano-scale. Eutectic Sn-Pb and near-eutectic Sn-Pb alloys are composed of a small number of eutectic colonies of lamellar Sn and Pb. Tin-based Pb-free alloys contain multiple phases that occur at multiple size scales depending on the local cooling conditions during the solidification process and on subsequent aging and deformation conditions as seen for a SAC alloy in Figure 2.22. For Sn-3.5Ag, the eutectic Sn-Ag alloy composition examined in detail by Chawla et al., cooling from above the eutectic temperature first results in the formation of Ag_3Sn plates since the high Sn-alloys do not solidify with a classic eutectic structure as a result of the difficulty in nucleating the β-Sn phase as reviewed in Section 2.4.1. [56-58] The size of the platelets depends on the cooling rate. Once the β-Sn phase is able to nucleate, β-Sn dendrites form with interdendritic Ag_3Sn of an intermediate scale and sub-100nm Ag_3Sn particles within the dendrites, with the size of all features depending on cooling rate.

Kerr and Chawla have elucidated the role, for example, of sub-100 nm Ag_3Sn precipitates on creep and grain boundary and dislocation motion in Sn-3.5Ag, as seen in Figure 2.31. [56] Figure 2.31a-c shows Ag_3Sn precipitates in β-Sn that pin the grain boundary in the as-solidified state and following deformation. Deformation produces dislocations whose motion is also pinned by the Ag_3Sn precipitates, as seen by the increased dislocation content and the intersections of dislocations and Ag_3Sn precipitates. For alloys containing Cu or Ni, either because it was in the initial alloy or

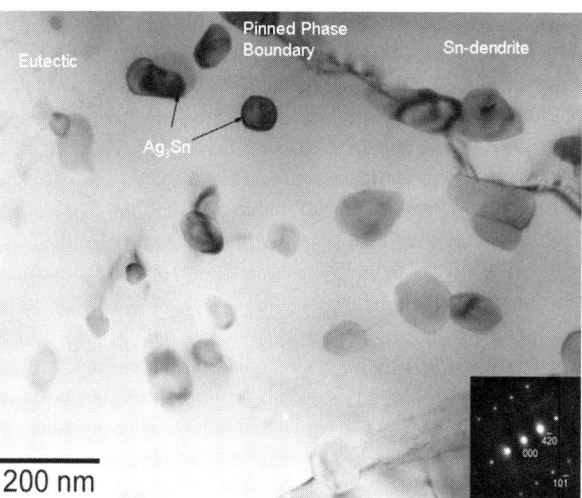

Fig. 2.31a. TEM micrographs from Kerr and Chawla showing microstructures of Sn-3.5Ag as solidified with the eutectic mixture containing the Ag_3Sn particles, Sn-dendrite, and the pinned boundary between the two phases. [56]

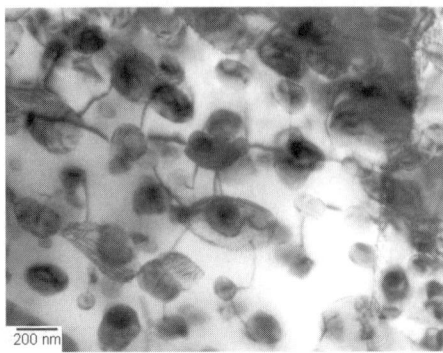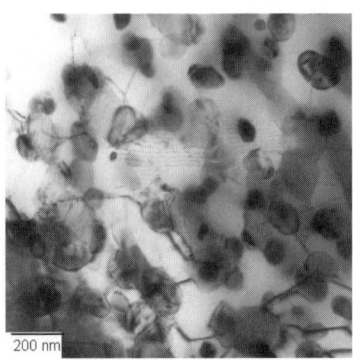

Fig. 2.31b, c. TEM micrographs from Kerr and Chawla showing microstructures of Sn-3.5Ag after creep deformation (22 MPa at 60°C, with a strain rate on the order of 10^{-4}/s, <110> zone axis): (b) Dislocations pinned by Ag_3Sn particles in the eutectic mixture after creep deformation (c) Pinning of dislocation substructure by Ag_3Sn particles. [56]

due to dissolution of the surface finishes, pad, and lead material into the solder, the structures will also contain additional intermetallic compounds whose size, scale and location depend on cooling rate.

Coarsening of these heterogeneous microstructures then occurs at room temperature and above for Pb-free alloys as well as for eutectic and near-eutectic Sn-Pb. [55] This leads to differences in mechanical properties for a given alloy solder joint with cooling rate, time after solidification, and thermal and stress history. [56] This contributes to the differences observed between seemingly "identical" experiments and between solder joints of the same alloy tested under different conditions.

In general, high-Sn Pb-free solders are stronger and have a higher creep resistance than Sn-Pb eutectic and near eutectic alloys. The creep data from Frear comparing the behavior of different solder alloys shown in Figure 2.32 [59] and yield strength data from the IDEALS project [11-14] as a function of composition and temperature shown in Figure 2.33 illustrate the major difference between Sn-Pb and high-Sn lead-free alloys. The dependence of the relative creep behavior of solder alloys on temperature, composition, and applied stress even in simple testing conditions is illustrated in the research of Guo et al. described in Table 2.4. [60] The differences seen here also show how changes in alloy composition resulting from reaction with the surface finishes or lead and substrate materials further change the alloy performance. Additional analyses by Guo et al. suggested that annealing the samples can have a profound effect on the creep behavior: the lower creep resistance seen in the Darveaux data in Figure 2.34 was attributed to the pre-annealing treatment applied to the samples tested by Darveaux in contrast to the Guo samples that were tested within one week of fabrication. [60-61]

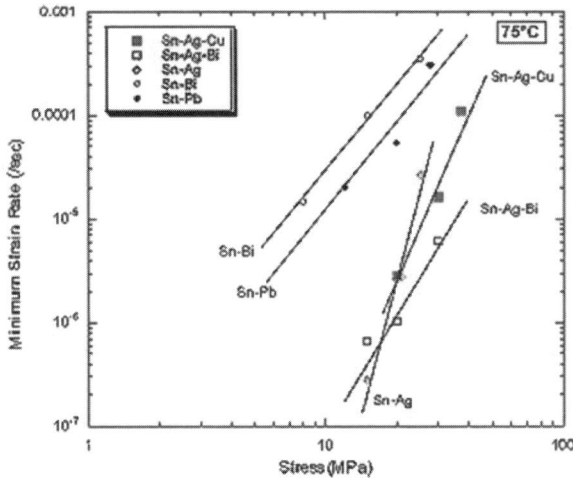

Fig. 2.32. Creep data for Sn-Pb eutectic, Sn-Bi eutectic and other compositions at 75°C as a function of applied stress. [59]

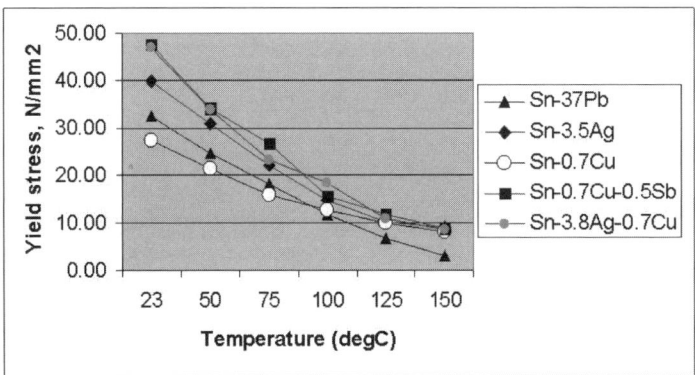

Fig. 2.33. Yield stress as a function of composition and temperature in °C. [11-14]

Table 2.4. Ranking of creep resistance of solder alloys based on best-fit power law creep by Guo et al. [60]

	Approx. 10 MPa low stress rank	Approx. 20 MPa high stress rank
Room Temperature		
Least creep resistant	Sn-2Ag-1Cu-1Ni	Sn-4Ag-0.5Cu
to	Eutectic Sn-3.5Ag	Sn-3.5Ag-0.5Ni
Most creep resistant	Sn-4Ag-0.5Cu	Sn-2Ag-1Cu-1Ni
	Sn-3.5Ag-0.5Ni	Eutectic Sn-3.5Ag
85°C		
Least creep resistant	Sn-2Ag-1Cu-1Ni	Sn-2Ag-1Cu-1Ni
to	Sn-4Ag-0.5Cu	Sn-3.5Ag-0.5Ni
Most creep resistant	Eutectic Sn-3.5Ag	Sn-4Ag-0.5Cu
	Sn-3.5Ag-0.5Ni	Eutectic Sn-3.5Ag

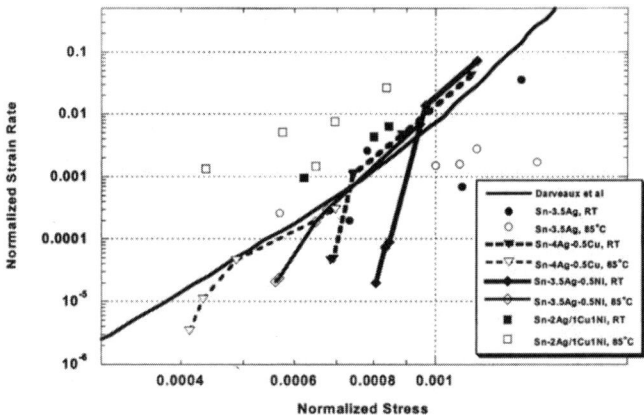

Fig. 2.34. Normalized steady-state creep strain rates versus normalized stresses for eutectic Sn-3.5Ag, Sn-4Ag-0.5Cu, Sn-2Ag-1Cu-1Ni and Sn-3.5Ag-0.5Ni solder joints along with Darveaux's data for aged eutectic Sn-3.5Ag solder joints. From Guo et al. [60]

Similarly careful experiments, performed to determine the dominant mechanisms for deformation during thermo-mechanical fatigue or creep, have come up with opposite results regarding the dominant mechanisms. For example, on the importance of grain boundary sliding to creep in Sn-3.5Ag, particularly with respect to the role of low angle boundaries. [56, 60] These differences are likely the result of microstructural or sample history differences outlined previously. What this suggests, therefore, is that general trends alone should be used to compare alloys: the variability in mechanical performance due to all the factors described, many times due to uncontrollable factors, leads to a large uncertainty in the performance of any solder joint.

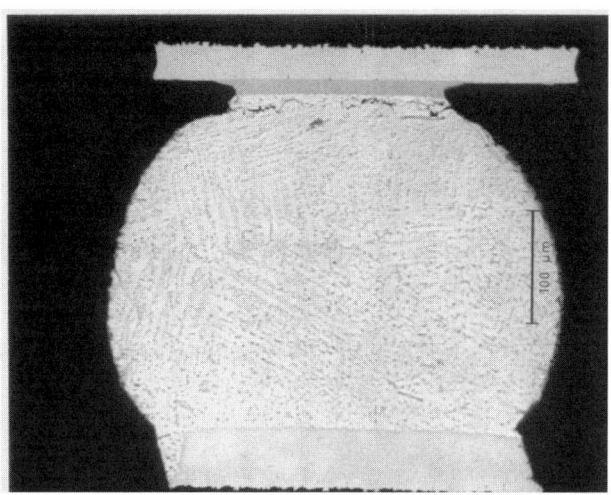

Fig. 2.35a. NEMI Lead-Free Assembly Project cross sections of 169 CSP Solder Joints fabricated with Sn-3.9Ag-0.6Cu after ATC: 0°C to 100°C. [18]

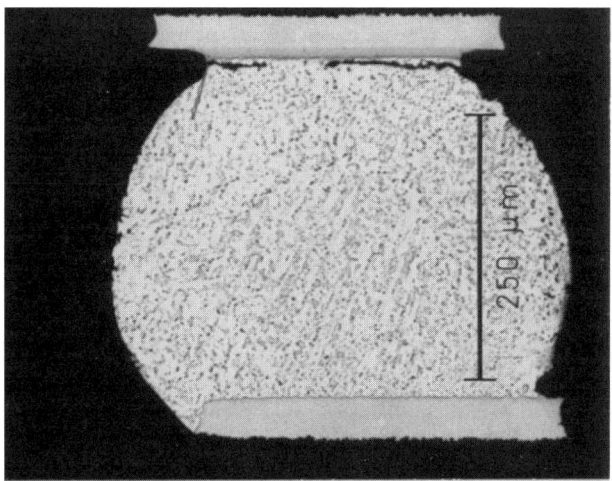

Fig. 2.35b. NEMI Lead-Free Assembly Project cross section of 169 CSP solder joint fabricated with Sn-3.9Ag-0.6Cu after ATC: –40°C to 125°C. Note that the microstructure in Fig. 2.35b is significantly coarser than Fig. 2.35a. [18]

The situation is even more complicated for accelerated thermo-mechanical cycling (ATC): during ATC, a solder joint will develop a changing composition, stress and microstructural state that varies spatially across the solder joint, resulting in changes with the number of thermal cycles, and varies with component type and position of the solder joint on the component. Typical CSP169 solder joints from the iNEMI Lead-Free Solder Project that underwent ATC at two different thermal cycling conditions are seen in Figure 2.35. The microstructures have coarsened differently as a result of the different thermal cycling conditions. [18]

2.7.2 Accelerated Thermal Cycling and Acceleration Factors

For Sn-Pb solders, there has been widespread acceptance of ATC of PWB test vehicles under specific conditions as a reliable method to assess thermo-mechanical fatigue resistance. This acceptance came from decades of industry experience relating specific thermal cycling conditions to wear-out failure in specific classes of electronic products. The microelectronics industry is just beginning to establish the practical experience relating design, product use, and ATC that took over fifty years for Sn-Pb assemblies.

The complexity of the situation with Pb-free solders was evident from the NCMS Pb-free Project in which the Pb-free solder alloys were observed to withstand different amounts, types, and rates of loading at different temperatures which are dependent upon the different coefficients of thermal expansion (CTE) and mechanical properties of the board, components, and alloys, solder joint geometry, solder microstructure, and residual stresses. [1-4] Taken together for a given alloy, these properties produce solder joint performance better for some components than eutectic Sn-Pb and worse for other components on the same board, and are likely to be different for different thermal cycling conditions.

A major remaining question is what are the proper acceleration factors for Pb-free solders, which quantify the relationship between solder composition, thermal cycling conditions, component and board materials and geometries, and useful assembly life. The current state of understanding was enunciated by Bartelo et al., in their examination of the relative performance of Sn-Pb eutectic and Pb-free solders as a function of ATC cycling conditions for a single component type, ceramic ball grid arrays (CBGA). [62] For 0°C to 100°C ATC testing with cycles times of 30 to 240 minutes, the ATC performances of CBGAs assembled with Sn-3.8Ag-0.7Cu and Sn-3.5Ag-3.0Bi (SAB) alloys were superior to those assembled with eutectic Sn-Pb CBGAs. When the thermal cycling condition was changed to –40°C to 125°C at cycle times from 42 minutes to 240 minutes, the ATC performance of CBGAs assembled with the SAC alloy was inferior to eutectic Sn-Pb CBGAs. Using the same temperature difference as 0°C to 100°C, the performance of the SAC alloy was superior to those assembled with eutectic Sn-Pb CBGAs. Similar results were obtained for the Sn-3.5Ag-3Bi alloy, with the added complication that for –40°C to 125°C ATC test conditions, the CBGAs assembled with SAB performed better than the Sn-Pb eutectic controls for short cycle times (42 minutes), but performed worse at longer ATC cycle times (240 minutes).

These results are similar to those obtained by Swan et al. for a wider variety of components, but a smaller set of cycling conditions. [63] Considerable data on the lifetimes of solder joints in commercial products coupled with extensive ATC testing are needed before acceleration factors can begin to be quantified for even a single alloy.

Future Work

Future research on Pb-free alloys must focus on the following issues:

- high reliability applications – server, military, aerospace, medical, for which exemptions are going away or where the lack of availability of tin-lead components will force a switch to lead-free soldering
- understanding whether there are any significant differences between different SAC alloy compositions in terms of reliability
- acceleration factors for SnAgCu and SnCu based alloys
- interactions of lead-free solder alloys and board/component surface finishes (containing gold, nickel, bismuth, silver, copper)
- effects of lead-free soldering time above liquidus temperature and peak soldering temperature on intermetallic formation and solder joint microstructure and its effect on reliability
- effect of lead-free cooling rate and subsequent aging on intermetallic compound and solder joint microstructure and their effect on reliability
- copper dissolution studies with different lead-free solder joint alloys and reliability of the resulting solder joint
- effect of CuSn, NiSn, NiCuSn, AgSn, AuSn intermetallic on lead-free solder joint reliability
- effect of lead-free surface mount, wave and rework (BGA, Hand, wave) processing on intermetallic formation and solder joint microstructure with the resultant affect on reliability
- effect of lead-free wave solder holefill on reliability

- effect of ATC and mechanical testing (bend, shock, vibration) on lead-free solder joint reliability
- effect of ATC dwell time and temperature ranges on lead-free solder joint reliability

Acknowledgements

The authors gratefully acknowledge the valuable contributions of their colleagues at NIST and their industry colleagues, including Jasbir Bath, Edwin Bradley, Srinivas Chada, Nikhilesh Chawla, Jean-Paul Clech, Darrel Frear, Polina Snugovsky, and Werner Engelmeier.

References

1. NCMS Lead-Free Solder Project Final Report, NCMS, National Center for Manufacturing Sciences, 3025 Boardwalk, Ann Arbor, Michigan 48108-3266, Report 0401RE96, August 1997, and CD-ROM database of complete dataset, including micrographs and raw data, August, 1999. Information on how to order these can be obtained from: http://www.ncms.org/
2. I. Artaki, D. Noctor, C. Desantis, W. Desaulnier, L. Felton, M. Palmer, J. Felty, J. Greaves, C.A. Handwerker, J. Mather, S. Schroeder, D. Napp, T.Y. Pan, J. Rosser, P. Vianco, G. Whitten, and Y. Zhu. Research trends in lead-free soldering in the US: NCMS lead-free solder project (Keynote). 602-605. 1999. IEEE Computer Society. Proceedings-EcoDesign '99: First International Symposium on Environmentally Conscious Design and Inverse Manufacturing; February 1-3, 1999; Tokyo, Japan.
3. C.A. Handwerker, E.E. de Kluizenaar, K. Suganuma, F.W. Gayle, Major International Lead-Free Solder Studies, Eds., Puttlitz K.J., Stalter K.A. Handbook of Lead-Free Solder Technology and Microelectronic Assemblies. New York: Marcel Dekker. In press.
4. C.A. Handwerker, NCMS lead-free solder project: a summary of results, conclusions and recommendations. 1999. IPC. IPC Work '99: An International Summit on Lead-Free Electronics Assemblies; Proceedings; October 23-28, 1999; Minneapolis, MN.
5. F. Gayle, G. Becka, A. Syed, J. Badgett, G. Whitten, T.-Y. Pan, A. Grusd, B. Bauer, R. Lathrop, J. Slattery, I. Anderson, J. Foley, A. Gickler, D. Napp, J. Mather, C. Olson, High temperature lead-free solder for microelectronics J. Minerals, Metals & Materials Soc. 53 (6): 17-21 June 2001.
6. F.W. Gayle, Fatigue-Resistant, High Temperature Solder, Advanced Materials & Processes 159 (4), April 2001, p. 43-44.
7. F.W. Gayle, G. Becka, A. Syed, J. Badgett, G. Whitten, T.-Y. Pan, A. Grusd, B. Bauer, R. Lathrop, J. Slattery, I. Anderson, J. Foley, A. Gickler, D. Napp,

J. Mather, C. Olson, High temperature lead-free solder for microelectronics, NCMS, Ann Arbor, MI, 2001 (published on CD-ROM).

8. J.H. Vincent, B.P. Richards, D.R. Wallis, I. Gunter, M. Warwick, H.A.H. Steen, P.G. Harris, M.A. Whitmore, S. Billington, A.C. Harman, and E. Knight, "Alternative solders for electronics assemblies. Part 2: UK progress and preliminary trials," Circuit World, 19: pp.32-34, 1993.

9. Alternative Solders for Electronic Assemblies – Final Report of DTI Project 1991-1993, GEC Marconi, ITRI, BNR Europe, and Multicore Solders. DTI Report MS/20073, issued 10.26.93.

10. H. Vincent and G. Humpston, "Lead-free solders for electronic assembly," GEC Journal of Research, 11: pp.76-89, 1994.

11. M. Harrison and J.H. Vincent, "Improved Design Life and Environmentally Aware Manufacturing of Electronic Assemblies by Lead-Free Soldering," http://www.lead-free.org/research/index.html.

12. M.R. Harrison and J. Vincent, "IDEALS: Improved design life and environmentally aware manufacturing of electronics assemblies by lead-free soldering", Proc. IMAPS Europe '99, (Harrogate, GB), June 1999.

13. C.A. Handwerker, F.W. Gayle, E. de Kluizenar, K. Suganama, Major International Lead (Pb)- Free Solder Studies, in Handbook of Lead-free Solder Technology for Microelectronics Assemblies, Eds. K.J. Puttlitz and K.A. Stalter, Marcel Dekker, 2004.

14. M.H. Biglari, M. Oddy, M.A. Oud, P. Davis, E.E. de Kluizenaar, P. Langeveld, and D. Schwarzbach, "Pb-free solders based on SnAgCu, SnAgBi, SnAg, and SnCu, for wave soldering of electronic assemblies," Proc. Electronics Goes Green 2000+, (Berlin, Germany) September 2000.

15. J. Bath, C.A. Handwerker, E. Bradley, Research Update: Lead-Free Solder Alternatives, Circuits Assembly, May 2000, 31-40.

16. E. Bradley, NEMI Pb-free interconnect task group report. 1999. IPC Works '99: An International Summit on Lead-Free Electronics Assemblies; Proceedings; October 23-28, 1999; Minneapolis, MN.

17. A. Rae and C.A. Handwerker, Circuits Assembly, April 2004, 20-25.

18. E. Bradley, J. Bath, C.A. Handwerker, R.D. Parker, Lead-Free Electronics: iNEMI Projects Lead to Successful Manufacturing, 2007, in press.

19. K. Suganuma, Research and development for lead-free soldering in Japan. 1999. IPC. IPC Work '99: An International Summit on Lead-Free Electronics Assemblies; Proceedings; October 23-28, 1999; Minneapolis, MN.

20. JEITA Lead-free Roadmap 2002 for Commercialization of Lead-free Solder, September 2002, Lead-Free Soldering Roadmap Committee, Technical Standardization Committee on Electronics Assembly Technology, JEITA (Japan Electronics and Information Technology Industries Association).

21. NEDO Research and Development on Lead-Free Soldering, Report No.00-ki-17, JEIDA, Tokyo, Japan, 2000.

22. Lead-Free Soldering – An Analysis of the Current Status of Lead-Free Soldering, Report from the UK Department of Trade and Industry. Copies can be obtained from the ITRI website: http://www.lead-free.org/

23. Second European Lead-Free Soldering Technology Roadmap, February 2003 and Framework for an International Lead-Free Soldering Roadmap, December 2002, Soldertec, available at http://www.lead-free.org.

24. ASM Binary Alloy Phase Diagrams, Eds. T.B. Massalski, H. Okamoto, P.R. Subramanian, L. Kacprzak, ASM International, 1990.

25. U.R. Kattner, Phase diagrams for lead-free solder alloys, JOM-JOURNAL OF THE MINERALS METALS & MATERIALS SOCIETY, 54 (12): 45-51, 2002.

26. U.R. Kattner, C.A. Handwerker. Calculation of phase equilibria in candidate solder alloys. Zeitschrift fur Metallkunde 2001; 92(7): 740-746.

27. K.W. Moon, W.J. Boettinger, U.R. Kattner, F.S. Biancaniello, C.A. Handwerker. Experimental and thermodynamic assessment of Sn-Ag-Cu solder alloys. Journal of Electronic Materials 2000; 29(10): 1122-1136.

28. U.R. Kattner and W.J. Boettinger, "On the Sn-Bi-Ag ternary phase-diagram," Journal of Electronic Materials 23 (1994) 603-10.

29. K.W. Moon, W.J. Boettinger, U.R. Kattner, C.A. Handwerker, D.J. Lee, The effect of Pb contamination on the solidification behavior of Sn-Bi solders, J Elec. Mater. 30 (1): 45-52 (2001).

30. S. Chada, W. Laub, R.A. Fournelle, and D. Shangguan, Copper substrate dissolution in eutectic Sn-Ag Solder and its Effect on Microstructure, J. Electronic Mater. 29, 1214-1221 (2000).

31. T.M. Korhonen, P. Su, S.J. Hong, M.A. Korhonen, and C.-Y. Li, Reactions of Lead-Free Solders with CuNi Metallizations, J. Electronic Mater. 29 1194-1199 (2000).

32. D. Swenson, The effects of suppressed beta tin nucleation on the microstructural evolution of lead-free solder joints, J. Mater. Sci. Mater. Electronics, 18 39-54 (2007).

33. J.W. Jang and D. Frear, unpublished research.

34. JCAA/JG-PP No-Lead Solder Project: −55°C to +125°C Thermal Cycle Testing Final Project, Rockwell Collins, D. Hillman and R. Wilcoxon, May 28, 2006.

35. K.S. Kim, S.H. Huh, and K. Suganuma, Effects of cooling speed on microstructure and tensile properties of Sn-Ag-Cu alloys, Mater. Sci. Eng. A. 333 106-114 (2002).

36. D.W. Henderson, T. Gosselin, A. Sarkhel, S.K. Kang, W.K. Choi, D.Y. Shih, C. Goldsmith, and K.J. Puttlitz, Ag3Sn plate formation in the solidification of near ternary eutectic Sn-Ag-Cu alloys, J. Mater. Res., 17 2775-2778 (2002).

37. I.E. Anderson, J.C. Foley, B.A. Cook, J. Harringa, R.L. Terpstra, O. Unal, Alloying effects in near-eutectic Sn-Ag-Cu solder alloys for improved microstructural stability, J. Electron. Mater. 30 (9): 1050-1059 (2001).

38. I.E. Anderson, B.A. Cook, J. Harringa, R.L. Terpstra, Microstructural modifications and properties of Sn-Ag-Cu solder joints induced by alloying, J. Electron.Mater. 31, 1166-1174 (2002).

39. K.L. Buckmaster, J.J. Dziedzic, M.A. Masters, B.D. Poquette, G.W. Tormoen, D. Swenson, D.W. Henderson, T. Gosselin, S.K. Kang, D.Y. Shih and K.J. Puttlitz, Presented at the TMS 2003 Fall Meeting, Chicago, IL, November, 2003.

40. M.A. Dudek, R.S. Sidhu, and N. Chawla, Novel rare-earth-containing lead-free solders with enhanced ductility, J. Metals 58 (6) 57-62 (2006).

41. K. Suganuma, Microstructural features of lift-off phenomenon in through hole circuit soldered by Sn-Bi. Scripta Materialia 1998; 38(9):1333-1340.
42. W.J. Boettinger, C.A. Handwerker, B. Newbury, T.Y. Pan, J.M. Nicholson, Mechanism of fillet lifting in Sn-Bi alloys, J. Elec. Mater., 31 (5): 545-550 MAY 2002.
43. H. Takao, H. Hasegawa, Influence of alloy composition on fillet-lifting phenomenon in Sn-Ag-Bi alloys, J. Elec. Mater., 30(2001), 513-520.
44. P. Biocca, Solder Paste: What are the process requirements to achieve reliable lead-free wave soldering?, http://www.leadfreemagazine.com/pages/papers/Q_A_Kester.pdf.
45. P. Biocca, Reliable Lead-Free Wave Soldering and SMT Processes, http://ap.pennnet.com/Articles/Article_Display.cfm?Section=Articles&Subsection=Display&ARTICLE_ID=216210.
46. K.-W. Moon, U.R. Kattner, and C.A. Handwerker, The Effect of Bi Contamination on the Solidification Behavior of Sn-Pb Solder, J. Elec. Mater. Nov. 2006.
47. NIST Thermodynamic Database for Solder Alloys, http://www.metallurgy.nist.gov/phase/solder/
48. J.-O. Andersson, T. Helander, L. Höglund, P. Shi, and B. Sundman, Thermo-Calc and DICTRA, Computational Tool for Materials Science, Calphad 26 (2002) 273-312.
49. W.J. Boettinger, U.R. Kattner, S.R. Coriell, Y.A. Chang and B.A. Mueller, A Development of Multicomponent Solidification Micromodels Using a Thermodynamic Phase Diagram Data Base, in Modeling of Casting, Welding and Advanced Solidification Processes, VII, Eds. M. Cross and J. Campbell, TMS, Warrendale, PA, 1995, 649-656.
50. D.W.G. White, Surface tensions of Pb, Sn, and Pb-Sn alloys, Metallurgical Transactions, 2 (1971) 3067-3071.
51. I. Ohnuma, X.J. Liu, H. Ohtani, K. Anzai, R. Kainuma, K. Ishida, Development of thermodynamic database for micro-soldering alloys, Electronics Packaging Technology Conference, 2000. (EPTC 2000). Proceedings of 3rd Conference, 2000, 91-96.
52. I. Ohnuma, M. Miyashita, K. Anzai, X.J. Liu, H. Ohtani, R. Kainuma, Phase equilibria and the related properties of Sn-Ag-Cu based Pb-free solder alloys. J. Elec. Mater. 2000; 29(10): 1137-1144.
53. N.-C. Lee. Prospect of lead-free alternatives for reflow soldering. 1999. ICP. IPC Work '99: An International Summit on Lead-Free Electronics Assemblies; Proceedings; October 23-28, 1999; Minneapolis, MN.
54. E. Bradley and J. Hranisavljevic, ECTC, 50th Electronic Components & Technology Conference, IEEE Transactions on Electronics Packaging Manufacturing 24:4 (10/2001) 255-260.
55. J.P. Clech, Report to NIST on Review and Analysis of Lead-Free Solder Material Properties, 2002, http://www.metallurgy.nist.gov/solder/clech/Introduction.htm.
56. M. Kerr and N. Chawla, Creep deformation behavior of Sn-3.5Ag solder/Cu couple at small length scales, Acta Mater. 52 4527-4535 (2004).

57. X. Deng, N. Chawla, K.K. Chawla, and M. Koopman, Deformation behavior of (Cu,Ag)-Sn intermetallics by nanoindentation, Acta. Mater. 52 4291-4303 (2004).
58. F. Ochoa, X. Deng, and N. Chawla, Effects of cooling rate on creep behavior of a Sn-3.5Ag alloy, J. Electron. Mater. 33 1596-1607 (2004).
59. Personal Communication D. Frear and E. Bradley (2000).
60. F. Guo, S. Choi, K.N. Subramanian, T.R. Bieler, J.P. Lucas, A. Achari, M. Paruchuri, Evaluation of creep behavior of near-eutectic Sn-Ag solders containing small amounts of alloy additions, Mater. Sci. Eng. A 35: 190-199 (2003).
61. R. Darveaux and K. Banerji, IEEE Trans. Components, Hybrid and Manufacturing Technology 15 1013-1024 (1992).
62. J. Bartelo, S.R. Cain, D. Caletka, K. Darbha, T. Gosselin, D.W. Henderson, D. King, K. Knadle, A. Sarkhel, G. Thiel, C. Woychik, "Thermomechanical fatigue behavior of selected lead-free solders", Proceedings, IPC SMEMA Council APEX 2001, Paper # LF2-2.
63. G. Swan, A. Woosley, K. Simmons, T. Koschmieder, T.T. Chong, and L. Matsushita. Development of lead-(Pb) and halogen free peripheral leaded and PBGA components to meet MSL3 at 260°C peak reflow profile. 1, 121-126. 9-11-2000. VDE Verlag. Proceedings, Electronics Goes Green 2000+, September 11-13, 2000, Berlin, Germany.

Chapter 3: Lead-Free Surface Mount Assembly

Sundar Sethuraman, Solectron Corporation, Milpitas, California

3.1 Introduction

In general, the process flow for lead-free surface mount assembly process is similar to the conventional SnPb soldering process. Often the same equipment set used for SnPb can be used for lead-free reflow soldering. However, there are some differences that must be taken into account. The material set used for lead-free soldering is different and typically higher reflow temperatures are required. This chapter will review the different aspects of the surface mount assembly process with respect to lead-free solder and discuss its impact on design, equipment, process and materials.

3.2 Solder Paste Alloy

For many years, eutectic Sn37Pb solder has been the primary material for metallurgically joining electronic components to Printed Circuit Boards. With the need to find a suitable lead-free alternative, there has been an extensive industry-wide development effort to look at different solder alloy choices.

Solder alloy selection involves consideration of several critical factors such as toxicity, availability, cost, process compatibility and various performance metrics, such as wettability, surface tension, liquidus temperature, plastic range, and fatigue characteristics.

SnAgCu family of alloys are gaining wide acceptance within the electronics industry, as a viable replacement for eutectic SnPb solder. The SnAgCu alloys currently used for surface mount assembly process have compositions containing 3.0-4.0 wt % Ag and 0.5-0.9 wt % Cu with remainder tin. The assembly process is virtually identical for all the compositions within that range. They all have a small pasty range of about $< 3°C$ with the liquidus ranging between 217°C and 220°C. Currently Sn3.0Ag0.5Cu

(SAC305) and Sn3.8Ag0.7Cu (SAC387) are the main solder paste alloys used for lead-free surface mount assembly. The Sn4.0Ag0.5Cu (SAC405) and Sn3.0Ag0.5Cu (SAC305) alloys are mainly used for lead-free BGA spheres. Studies conducted by IPC Solder Products Value Council [3] showed no significant difference in reliability between the three different SnAgCu solder paste alloys (SAC305, SAC387 and SAC405) with the recommendation to use SAC305 as a general lead-free solder paste alloy replacement.

3.3 Screen Printing Process

The printing process for lead-free assembly is similar to that of the SnPb assembly process. No major changes should be necessary. Existing screen printers are likely to be useable without any modification. However, the use of lead-free solder paste may require some slight adjustments to the process such as modifying print pressure and print speed.

3.3.1 Stencil Design

Current stencil design and stencil fabrication methods used for SnPb solder paste printing would work for lead-free solder pastes. The stencil design guidelines specified in IPC-7525 [1] for SnPb solder paste should be applicable as well.

Lead-free solders have a higher surface tension and do not spread as well as the SnPb solder. So they tend to leave the edge of the PCB pads exposed after reflow. This may be more prominent on PCBs with OSP surface finish. To reduce this occurrence, the stencil apertures could be widened to allow more solder paste coverage on the pad maximizing solder spread during reflow. For example, apertures with 1-mil [0.0254mm] reduction (on all sides of the pad) used for SnPb solder paste could be replaced with 1:1 stencil aperture opening (with respect to PCB pad size) for lead-free solder pastes.

3.3.2 Printability

Lead-free SnAgCu solder pastes have similar printing characteristics as the SnPb solder paste. They exhibit similar performance in terms of print definition, print consistency and stencil life. As with SnPb solder paste, depositing adequate, consistent paste volume with good registration on the PCB pad is critical to achieving low defect rates.

For the solder paste to have good printability, the metal to binder volume ratio is critical. The flux accounts for nearly 50% of the solder paste by volume. The SnAgCu alloys have a slightly lower density (7.4gms/cc approx.) compared to the eutectic Sn37Pb solder (8.4gms/cc approx.). So in order to maintain the same metal to flux binder volume ratio, the metal content is generally reduced in SnAgCu solder pastes (approx. 88 wt % metal content compared to 90 wt % in SnPb solder paste).

The printability is also affected by the property of the flux binder. Most solder paste manufacturers have different flux formulation for their lead-free and SnPb solder pastes. The flux composition affects the rheological properties of the solder paste such as paste viscosity, tack life, stencil life, hot and cold slump characteristics. These properties need to be considered while selecting the appropriate lead-free solder paste. The print settings like print pressure and print speed need to be optimized for the selected solder paste to achieve good printing results.

In cases where both SnPb and lead-free assemblies are processed on the same assembly line, it is important to have dedicated squeegee blades, stencils and tooling (like spatula, board supports, fixtures) for each process to avoid lead contamination on lead-free assemblies.

3.3.3 Solder Paste Handling

The lead-free solder pastes may have shorter shelf life and more stringent storage conditions compared to the SnPb solder pastes. However, the general handling procedure should remain the same. It is critical to follow the handling guidelines of the solder paste manufacturer to achieve good results and avoid any major issues further down the process. It is also recommended to have separate storage areas for SnPb and lead-free solder pastes to avoid cross contamination. Solder paste suppliers typically take a cautious approach when recommending shelf life and usually specify a shorter shelf life for lead-free solder paste, especially with limited production experience.

3.4 Component Placement

Existing placement equipment should be able to place lead-free components without any issues. The difference in appearance of lead-free solder and lead-free components may require adjustments to the vision system for proper recognition.

3.5 Reflow Soldering

Conversion to lead-free solder has a huge impact on the reflow process with the higher melting point of the SnAgCu alloy offering the main challenge.

Most lead-free solder pastes that were initially available had the same flux formulation that was used for SnPb solder pastes, which lead to reflow issues such as excessive voids caused by the flux not being rated to the higher lead-free soldering temperatures. Since then there has been major development in the flux formulation for lead-free solder pastes. The newer generation solder pastes have fluxes with higher activation temperatures to account for the higher soak and reflow temperatures required for the lead-free solder.

3.5.1 Reflow Equipment

Most convection type reflow ovens used today are capable of achieving the higher temperatures necessary for the lead-free reflow process. Reflow equipment with an increased number of zones (typically about 10 heating zones) will have better temperature control to handle large complex server type assemblies. Ovens with fewer zones (4 to 5 heating zones) may be able to handle small low complexity cell phone type assemblies. But with more thermally challenging assemblies, they will have difficulty in minimizing the temperature delta across the board. Lowering the belt speed to achieve the desired peak temperatures may not be feasible as the PCB may have longer exposure at higher temperatures and the throughput would be affected.

Infrared ovens are generally not suited for lead-free reflow soldering. With infrared ovens, the amount of heating varies within the assembly depending on the surface texture and color of the parts reflowed. This typically leads to a large thermal gradient across the board.

Vapor phase reflow systems are now gaining some importance in the area of lead-free soldering. The main advantage of these systems compared to the convection reflow ovens, are their ability to maintain a low temperature delta across the board. The maximum temperature of the board does not exceed the boiling point of the medium (typically 230°C or 240°C) irrespective of the time it remains in vapor. However the infrastructure for vapor phase reflow equipment is not widely available, with only a small niche likely for extremely thick difficult to solder lead-free surface mount assemblies.

3.5.2 Thermal Profiling

Due to the narrow process window of lead-free solders, accurate thermal profiling is extremely important. As the component and PCB temperature ratings are now typically closer to the reflow temperatures, more locations have to be monitored. Existing thermal profilers could be used with minor modifications (such as adding a high heat resistant sleeve/cover). The thermocouples and thermally conductive adhesives used for profiling have to be rated for at least 260°C minimum.

For accurate monitoring of the solder joint and component body temperatures, the thermocouple attachment method needs to be reliable and consistent. Determining the locations for attaching thermocouples will need a careful review of the components on the PCB. Thermocouples should be attached to the solder joints of critical and heat sensitive components. If any lead-free component has a lower temperature rating (for example < 250°C), a thermocouple should be attached to the body of that component to ensure it does not exceed its temperature rating. Also, the best place to attach thermocouples would be the hottest and coldest solder joints on the board. This could be determined based on the thermal mass and location of the components on the board. Typically smaller devices like chip resistors will reach high temperatures and larger devices like high I/O BGAs and QFPs will have lower solder joint temperatures. It is also critical to monitor the peak temperature of the PCB to avoid any board damage due to excess heat.

3.5.3 Reflow Profile

Two types of reflow profiles are generally used (1) Ramp-Soak-Spike Profile and (2) Ramp-to-Peak Profile. The newer generation lead-free solder pastes are more suited towards Ramp-to-Peak reflow profiles with a relatively short soak time. The need for longer soak has reduced considerably as today's convection ovens have better heat transfer and temperature control compared to the older style infrared ovens.

The reflow profile for SnAgCu solder paste can be broken down into four stages.

1. **Preheat Stage:** The PCB and components are heated up gradually from room temperature to about 170°C. The volatile constituents in the flux evaporate. The critical parameter in the preheat stage is the ramp rate. It should be controlled between 1-3°C/sec. Higher ramp rates may cause thermal shock to the PCB and the components. Rapid heating could also result in flux splattering leading to solder balling.

2. **Soak Stage:** Also known as the dry-out or pre-reflow stage, the temperature is maintained between 170 to 220°C for an extended period, typically about 60 seconds for a Ramp-to-Peak profile and 120 seconds for a Ramp-Soak-Spike profile. This soak helps reduce the temperature delta between the large and small thermal mass components on the assembly during reflow. During this stage, the flux is activated and it cleans the surface oxides on the solder particles, component leads and PCB pads. Activated flux continues to keep the metal surfaces from re-oxidizing. Longer soak times could cause the fluxes to breakdown prior to reflow, leading to solderability issues.

3. **Reflow Stage:** Following the soak stage, the assembly enters the reflow stage where the solder particles melt and form a bond between the PCB and component terminations. Typically a peak temperature range of 235 to 250°C at the solder joint with time above liquidus ranging between 45-75 seconds should provide adequate time for wetting and formation of a quality solder joint. Higher peak temperatures (typically > 250°C) may result in flux charring and potential PCB/component damage. Longer reflow times (typically > 100 seconds) should be avoided as it can result in excessive Intermetallic (IMC) formation, potentially making the solder joint more brittle.

   Use of Nitrogen atmosphere during reflow helps reduce oxidation and enhances solder wetting. It aids solderability and improves the process window of lead-free solder pastes. However, the extent of improvement will depend on the solder paste flux chemistry. For PCBs with OSP surface finish, using nitrogen at reflow also helps in improving the barrel fill during the follow-on wave soldering process by reducing the oxidation of unsoldered pads and plated through vias.

4. **Cool down Stage:** During this stage, the assembly is cooled down to room temperature. A rapid cool down would be preferred which helps form a finer grain structure (typically 3 to 4°C/sec). Slow cooling will result in a larger grain size that may have poor fatigue resistance. However, excessive temperature gradients should be avoided to prevent potential damage to the components and PCB.

A typical reflow profile for lead-free SnAgCu solder paste is shown in Figure 3.1 with the eutectic Sn37Pb reflow profile shown as reference.

3.5.4 PCB and Component Warpage

There is an increased risk for warpage related issues with the higher temperature lead-free reflow process. Some of the high I/O area array devices,

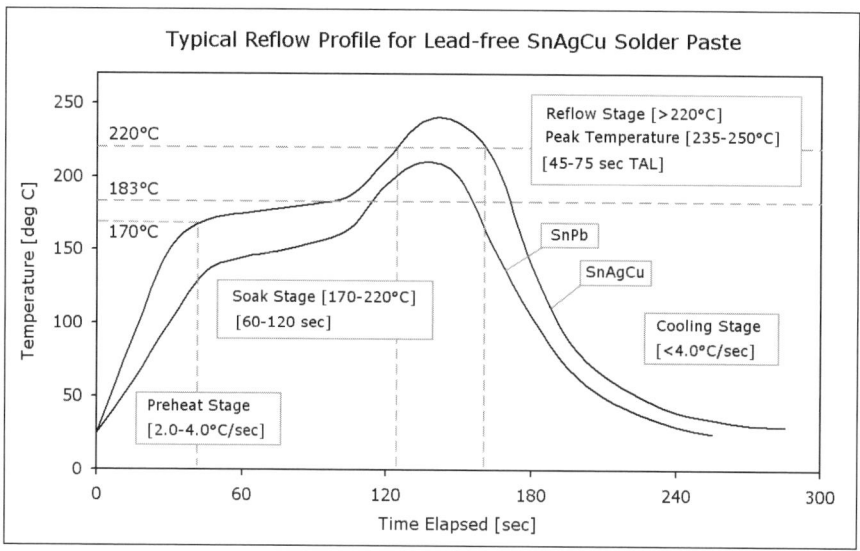

Fig. 3.1. Typical reflow profile for lead-free SnAgCu solder paste (eutectic SnPb reflow profile shown as reference)

BGA style sockets and long surface mount connectors may exhibit significant warpage during reflow which could lead to potential open solder or bridging defects.

There is also a potential for excessive warpage and delamination on PCBs during lead-free reflow, especially when FR-4 materials not rated to the higher lead-free soldering temperatures are used. So it is critical to keep the component body and PCB temperatures as low as possible in order to reduce warpage and avoid any heat related damage during reflow.

3.5.5 Flux Residues

Depending on the type of solder paste material used, the boards may or may not need cleaning. For lead-free water-soluble solder pastes, the wash process used for cleaning the flux typically remains the same. For no-clean solder pastes, the higher lead-free soldering temperatures may leave flux residues on the PCB that are harder to penetrate through with ICT bed of nails test probes. This could lead to false calls during test. Use of nitrogen atmosphere during reflow could help reduce the probing issues with lead-free solder paste. Post-reflow ageing has also shown to have an impact on probing with more false calls observed with longer ageing in certain cases.

3.5.6 Moisture Sensitivity Classification

As a result of higher reflow temperatures, there are further restrictions in handling moisture sensitive devices. As per IPC/JEDEC J-STD-020C [4], devices classified as Level 1 are not moisture sensitive and devices classified as Level 6 as extremely moisture sensitive. In order to maintain the MSL rating of SnPb components for their lead-free version, the component manufacturers may have to change the materials used in these devices or consider component redesigns. Typically for every 10°C rise in peak reflow temperature, the Moisture Sensitive Level (MSL) drops by one level.

The IPC/JEDEC J-STD-020C [4] specifies the temperatures at which MSL should be rated for lead-free plastic surface mount devices, based on package thickness and volume (refer to Table 3.1). It is also important to note that moisture sensitivity classification specified in this standard refers to the topside package body surface temperature rather than the solder joint temperature. Although the IPC/JEDEC J-STD-020C [4] is intended for plastic surface mount devices, many component manufacturers specify maximum temperatures for other device types using this standard.

The labeling and handling requirements for moisture sensitive plastic surface mount devices are defined in IPC/JEDEC J-STD-033B [5].

Table 3.1. Package Classification Reflow Temperatures for Lead-free Process specified in IPC/JEDEC J-STD-020C [4]

Package Thickness	Volume < 350 mm^3	Volume $350 - 2000$ mm^3	Volume > 2000 mm^3
< 1.6 mm	$260 + 0°C^*$	$260 + 0°C^*$	$260 + 0°C^*$
$1.6 - 2.5$ mm	$260 + 0°C^*$	$250 + 0°C^*$	$245 + 0°C^*$
≥ 2.5 mm	$250 + 0°C^*$	$245 + 0°C^*$	$245 + 0°C^*$

* The device manufacturer/supplier should assure process compatibility up to and including the stated classification temperature at the rated MSL level (generally rated for three reflow cycles). A lead-free component classified at a temperature below 260°C, should be capable of being reworked at 260°C within eight hours of removal from dry storage or bake, per IPC/JEDEC J-STD-033 [5].

3.6 Solder Joint Inspection and Acceptance Criteria

3.6.1 Visual Inspection

Lead-free solder joints generally have more surface roughness exhibiting a dull and grainy appearance. Due to the higher surface tension of lead-free alloys, greater wetting contact angles are formed (contact angle between solder to component and solder to PCB termination). This results in reduced spreading potentially exposing the base metal, especially with OSP PCB surface finishes. Figures 3.2 to 3.4 show some representative optical images of SnPb and SnAgCu soldered joints.

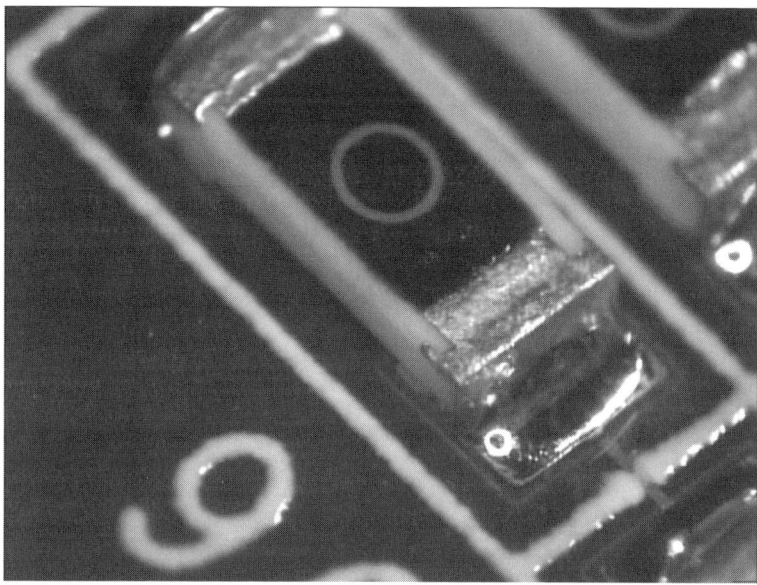

Fig. 3.2(a). Representative optical image of a 1206 chip resistor solder joint soldered with Sn37Pb (Sn10Pb coated resistor termination on the resistor and OSP over Cu on the PCB)

Fig. 3.2(b). Representative optical image of a 1206 chip resistor solder joint soldered with SnAgCu (Pure matte Sn coated termination on the resistor and OSP over Cu on the PCB)

Fig. 3.3(a). Representative optical image of SSOP solder joints soldered with Sn37Pb (Sn10Pb coated termination on the lead-frame and OSP over Cu on the PCB)

Fig. 3.3(b). Representative optical image of SSOP solder joints soldered with SnAgCu (Pure matte Sn coated termination on the lead and OSP over Cu on the PCB)

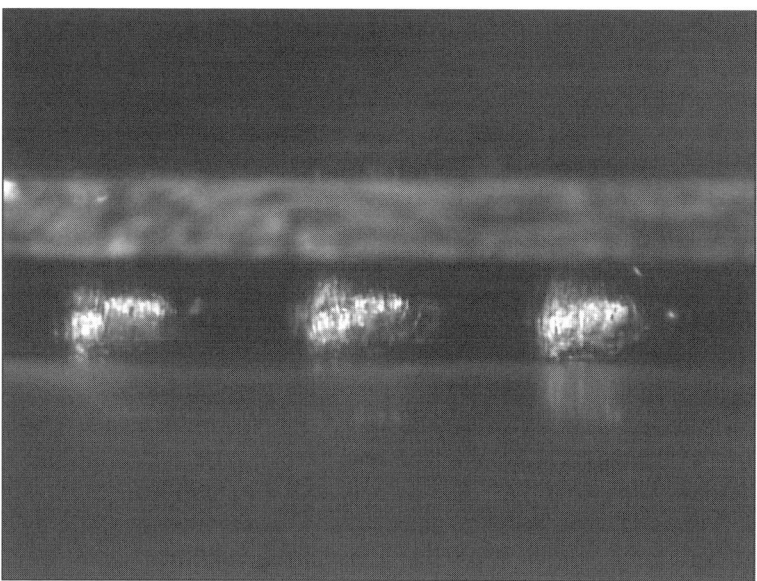

Fig. 3.4(a). Representative optical image of PBGA solder joints soldered with Sn37Pb (Sn37Pb spheres on the BGA and OSP over Cu on the PCB)

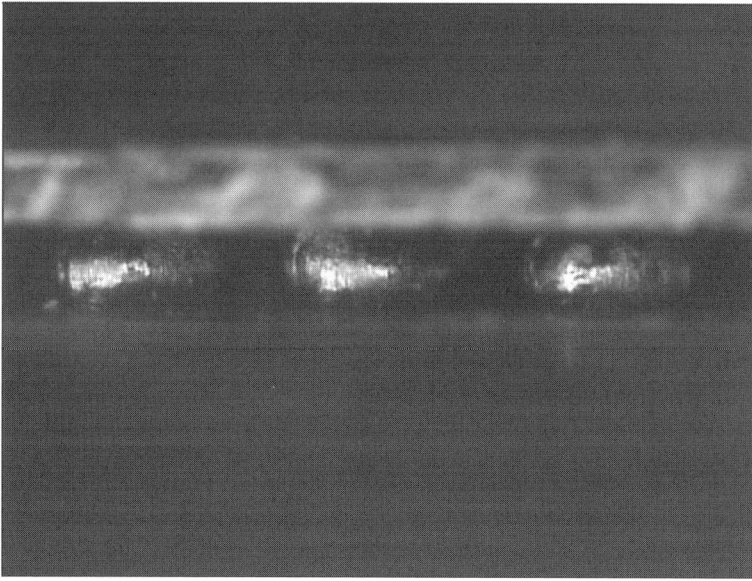

Fig. 3.4(b). Representative optical image of PBGA solder joints soldered with SnAgCu (SnAgCu spheres on the BGA and OSP over Cu on the PCB)

The acceptance criteria for lead-free alloys have been included in the latest revision of the IPC-A-610 document (Revision D) [2]. As per the document, dull and grainy joints are acceptable but non-wetting and de-wetting are not.

Generally for Automated Optical Inspection (AOI), the current inspection algorithms used for inspecting SnPb solder joints would work. Some reprogramming will be necessary due to differences in appearance of lead-free solder joints.

3.6.2 X-ray Inspection

The elimination of lead from the solder joint will result in X-ray images with relatively less contrast. However for inspection purposes, there is virtually no difference between the appearance of SnPb and lead-free SnAgCu solder joints under X-ray. So for Automated X-ray Inspection (AXI), the same inspection algorithms used for inspecting SnPb solder joints would work. The thresholds may need to be adjusted due to slight variations in grey scale values.

Figures 3.5 through 3.7 show some representative X-ray images of solder joints soldered with SnPb and SnAgCu. It can be observed from Figures 3.7(a) and 3.7(b), that solder joints soldered with lead-free SnAgCu have slightly more voiding than the solder joints soldered with SnPb. Even though these voids are not considered a reliability concern, the inspection algorithms may have difficulty in processing them leading to more frequent false fails. So they need to have some form of void compensation option to reduce false fails when inspecting lead-frame solder joints soldered with lead-free SnAgCu.

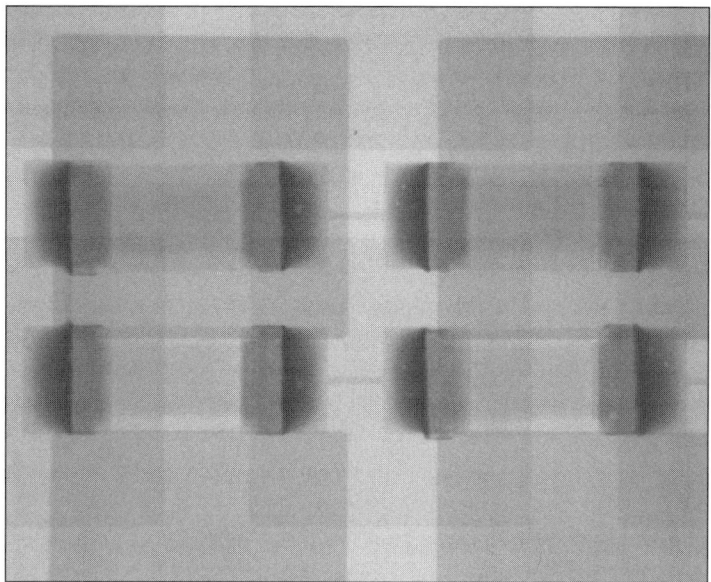

Fig. 3.5(a). Representative X-ray image of 1206 chip resistor solder joints soldered with Sn37Pb (Sn10Pb coated termination on the resistor and OSP over Cu on the PCB)

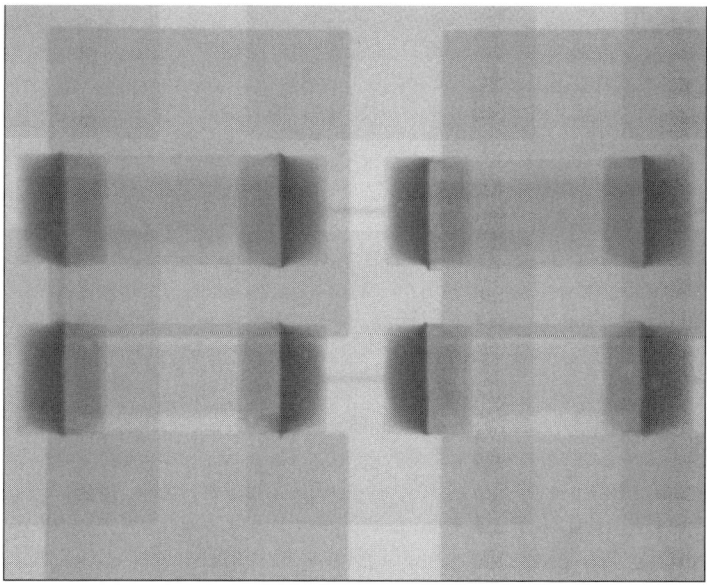

Fig. 3.5(b). Representative X-ray image of 1206 chip resistor solder joints soldered with SnAgCu (Pure matte Sn coated termination on the resistor and OSP over Cu on the PCB)

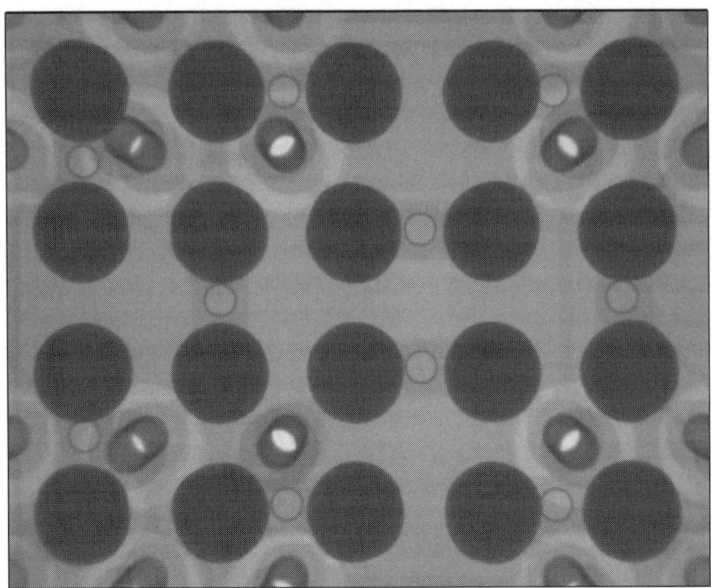

Fig. 3.6(a). Representative X-ray image of PBGA solder joints soldered with Sn37Pb (Sn37Pb spheres on the BGA and OSP over Cu on the PCB)

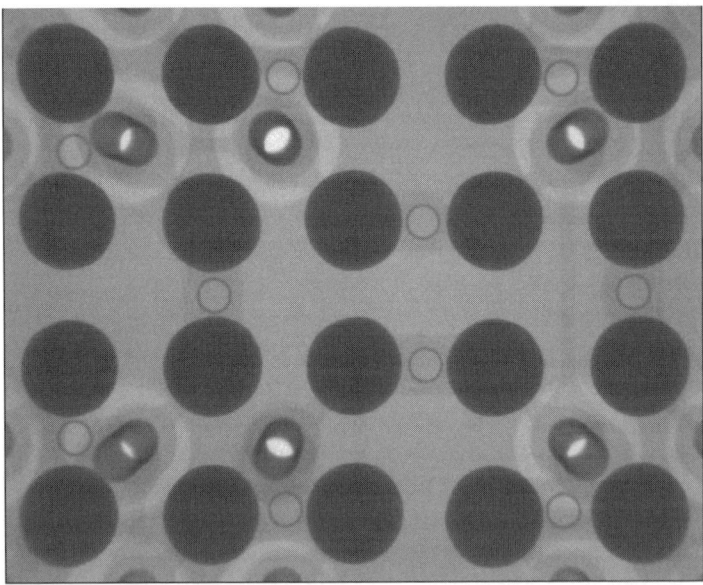

Fig. 3.6(b). Representative X-ray image of PBGA solder joints soldered with SnAgCu (SnAgCu spheres on the BGA and OSP over Cu on the PCB)

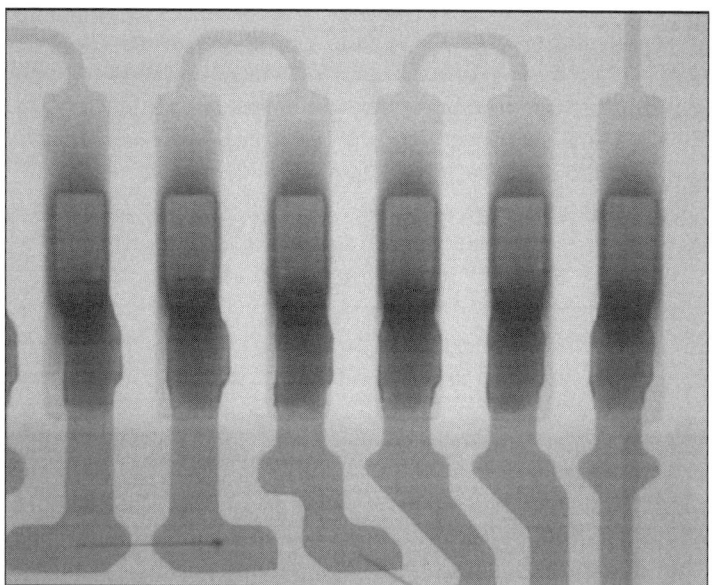

Fig. 3.7(a). Representative X-ray image of SSOP solder joints soldered with Sn37Pb (Sn10Pb coated termination on the lead and OSP over Cu on the PCB)

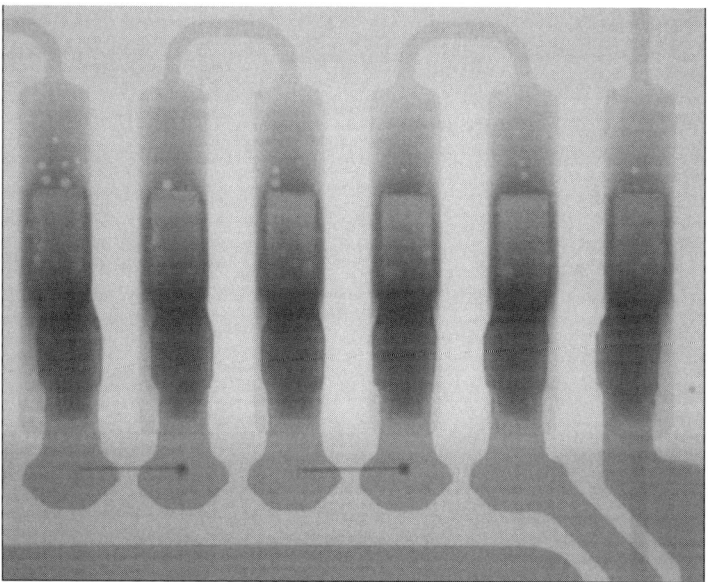

Fig. 3.7(b). Representative X-ray image of SSOP solder joints soldered with SnAgCu (Pure matte Sn coated termination on the lead and OSP over Cu on the PCB)

Summary

The printability of lead-free SnAgCu solder pastes have been shown to be equivalent to that of SnPb in most cases. There have been no placement issues reported with lead-free components.

The major difference between the SnPb and lead-free surface mount assembly process is the increase in reflow temperature. Reflow equipment with an increased number of zones may be needed for more thermally challenging assemblies. Due to the narrow process window of lead-free solders, accurate thermal profiling becomes critical. With the high temperature lead-free reflow, there is an increased risk of component and PCB warpage issues. In order to reduce this risk, materials rated for lead-free processing temperatures should be used.

The wash process after soldering remains the same for lead-free water soluble solder pastes. More probing issues may be experienced with no-clean fluxes due to the high temperature reflow making the residue more harder.

The visual inspection acceptance criteria for lead-free solder alloys have been updated to accommodate for the differences observed in the solder joint appearance. Some reprogramming may be necessary for AOI inspection. There are no major differences between the appearance of SnPb and lead-free SnAgCu solder joints under X-ray. The same inspection algorithms could be used. Due to slight variations in the grey scale values, the thresholds may need to be adjusted.

References

1. IPC-7525 (2000) Stencil Design Guidelines. IPC, Bannockburn, Illinois
2. IPC-A-610D (2005) Acceptability of Electronic Assemblies. IPC, Bannockburn, Illinois
3. IPC Round Robin Testing and Analysis of Lead Free Solder Pastes with Alloys of Tin, Silver and Copper. IPC Solder Products Value Council, Bannockburn, Illinois
4. IPC/JEDEC J-STD-020C (2004) Moisture/Reflow Sensitivity Classification for Nonhermetic Solid State Surface Mount Devices. JEDEC Solid State Technology Association, Arlington, Virginia
5. IPC/JEDEC J-STD-033B (2005) Handling, Packing, Shipping and Use of Moisture/Reflow Sensitive Surface Mount Devices. JEDEC Solid State Technology Association, Arlington, Virginia

Chapter 4: Lead-Free Wave Soldering

Christiane Faure, Solectron Corporation, Bordeaux, Cedex, France.
Jasbir Bath, Solectron Corporation, Milpitas, California, USA.

4.1 Introduction

Lead-free wave soldering is already in mass production in Asia and developing in Eastern Europe and the Americas, however lead-free wave is still the most challenging process to implement (together with lead-free BGA/CSP and PTH rework) and manufacturers around the world are challenged to qualify equipment and processes that will produce a quality lead-free product.

Lead-free alloys behave differently than SnPb alloys. The typical differences are:

- Lead-free alloys have slower wetting times than SnPb
- Lead-free alloys are more erosive (to board and component materials and equipment)
- Bridging is increased with lead-free solder
- Lead-free solder joints typically have a different solder joint appearance
- Insufficient hole fill can be more of an issue for lead-free wave soldering.

Typically, it should be always kept in mind that the process window for lead-free alloys are narrower than that of tin-lead solder and due to the higher solder melting point, lead-free alloys operate closer to the component and board working temperature limits. Selecting the right materials to make the wave equipment compliant for lead-free processing, includes the lead-free alloy, flux and process parameters and additional issues such as avoiding cross-contamination in a manufacturing plant assembling both tin-lead and lead-free products. It should also be mentioned that new inspection criteria for solder joints need to be set-up. Implementing lead-free wave soldering involves more than replacing one solder alloy with another and no unique solution exists.

4.2 Lead-Free Wave Solder Alloy Alternatives

4.2.1 Solder Alloys

The following section contains alloys that have been considered as candidates for replacing eutectic SnPb systems.

4.2.1.1 Sn3.5Ag

There is long experience with using this alloy. This eutectic composition has a melting point of 221°C. It was not typically widely considered for wave soldering as it produces a high-viscosity oxide film on the solder wave, which affects the flow properties of the solder [1]. Assemblers recommend this alloy as a good lead-free solder wire alternative for manual rework touch-up.

4.2.1.2 Sn3.0-4.0Ag0.5-0.9Cu

Tin-silver-copper (SAC) alloys used for lead-free soldering have composition ranging from 3.0 to 4.0% weight for silver and 0.5 to 0.9% weight for copper. The eutectic temperature of this ternary alloy is about 217°C where the exact composition has been demonstrated to be with 3.5±0.3%wt Ag and 0.9±0.2%wt Cu [2].

Copper was added to SnAg to slow Cu dissolution, lower melting temperature and improve wettability, creep and thermal fatigue characteristics (3). Today, for lead-free wave soldering, SnAgCu with 3.0%wt Ag and 0.5%wt Cu (called SAC305) is the most widely used tin-silver-copper composition used for lead-free wave soldering, due to cost considerations, compared to the higher silver composition (Sn4.0wt%Ag0.5wt%Cu).

In Japan, the tin-silver-copper alloys have been more widely used than tin-copper in wave soldering and more companies in Europe and the Americas have made SnAgCu alloys their preferred choice for wave soldering, even though they are more expensive than tin-copper.

4.2.1.3 Sn0.7Cu

This eutectic composition has a melting point equal to 227°C. The primary interest is the lower cost compared with silver containing alloys. The counterpoint of this alloy, well reported in the literature, is the lack of fluidity that results in the poorer through hole filling and increased solder bridging between component leads. The reduced wetting forces/times of SnCu alloy are translated in production into longer contact times at the wave to achieve equivalent hole fill compared with SnAgCu.

4.2.1.4 Sn0.6Cu0.05Ni (SN100C)

A small amount of nickel into SnCu solder has been found to improve the fluidity of SnCu alloy. Due to the small amount of nickel added, the eutectic melting point of the alloy is kept at 227°C. This formulation has been patented by Nihon Superior and sold under the name SN100C [4]. Another interest of this nickel modified SnCu alloy is that solder joints appear bright and are difficult to distinguish visually with SnPb solder; even with low cooling rates. It has also been reported that SnCuNi alloy is less erosive to materials (PCB copper traces and pads) compared to SAC alloys [4].

Although SnAgCu and Sn0.7Cu alloys were in 2003 the most popular ones for wave soldering [5], SnCuNi would appear today to be a real challenger with a good compromise between cost and quality. The patent and licensing implications are the main drawbacks to the widespread use of this alloy as is the melting point of SnCuNi which is 10°C higher than SnAgCu which would be a concern for lead-free hole fill considerations for thick PCBs.

4.2.1.5 Sn2.5Ag0.8Cu0.5Sb (CASTIN)

Sn2.5Ag0.8Cu0.5Sb, termed as CASTIN which is patented by AIM Solder, has a melting point in the 217°C range. It is suspected that the addition of Antimony may reduce copper-tin intermetallic growth and it is known that the antimony helps to improve the thermal fatigue resistance of the alloy [3,6]. It is recognized to have reduced copper erosion on PCB copper pads or traces compared to SnAgCu alloys.

4.2.1.6 SACX

Vaculoy SACX0307 is a lead-free alloy proposed by Alpha Metals as a replacement for Sn37Pb alloy in the wave solder process. The composition is Sn0.3Ag0.7CuX. X means that the product has been enhanced with the addition of 2 minor elements with the objective to reduce dross formation and improve joint cosmetics. The alloy has a melting point around 227°C [7].

Other alloys such as tin-silver-bismuth-copper are also lead-free alternatives for wave soldering, but today most companies prefer the concept of typically having the same alloy for reflow soldering and wave soldering and are adopting SnAgCu alloys for both.

4.2.2 Solder Bar Cost

The cost of the solder bar is directly dictated by that of the raw materials. Tin remains the base element in lead-free solder and the cost of solder bar has a tendency to increase firstly because of increased percentage and demand for tin. A major factor for cost increase is the presence of precious metals like silver. But the price of solder bar is also dictated by its shape (standard ingot bar compared to custom shape for automatic feeding of the equipment) and by elemental impurity levels. Finally, patent/licensing on alloy compositions may have an impact on the solder bar cost.

Table 4.1 gives a relative comparison between the mentioned alloys based on average cost on March and April of 2006 and an order of 1000 Kg from paste/alloy manufacturers data (AIM/Cookson/Kester) using the LME (London Metal Exchange) pricing.

Table 4.1. Cost comparison of tin-lead solder bar with different lead-free solder bar alloys

Solder Alloy	Relative Cost
Sn37Pb	1
Sn3Ag0.5Cu	3.3
Sn3.8Ag0.7Cu	3.9
SACX	1.9
Sn0.7Cu	1.6
Sn0.6Cu0.05Ni (SN100C)	2.1
Sn2.5Ag0.8Cu0.5Sb (Castin)	2.4

Basically, all the above alloys are suitable for the wave soldering process. But implementing in real life appears more complicated than for tin-lead and selection of the alloy has to be based on the final product application, cost and ease of implementation considerations.

4.3 Wave Solder Equipment Recommendations

4.3.1 Equipment Upgrade

Among all recommendations, it is important to understand that because lead-free alloys are less dense than tin-lead alloys, and the hardware used with wave equipment will sink to the bottom of the pot if dropped into the molten lead-free solder [8].

Many lead-free alloys cause erosion to the base metals used for solder pots, impellers and baffles because of the combination of the aggressiveness and higher percentage of tin and the high temperature. The surface of many base metals such as stainless steel generally show signs of pitting and start to dissolve after prolonged contact with lead-free high tin solder alloys. This leaching process releases iron particles, resulting in iron additions to the solder alloy. Without protection, parts made with stainless steel which are in contact with the lead-free alloy have proven to be eroded even after a few months use as shown in Figure 4.1.

All machine parts that are in contact with the molten alloy may need to be retrofitted. The pot is not the only part to consider to retrofit but the nozzles, flow ducts, conveyor fingers and pump hardware. Various equipment manufacturers have proposed a variety of solutions. The installed base of wave soldering machines is large and it is important to understand the impact to the equipment before switching to lead-free solder. Some machines may require very little changes regarding materials, while other machines may require replacement of the entire soldering unit. There are many varied solutions to prevent the degradation and erosion. The known alternatives being offered are cast gray iron, ceramic-coated stainless steel, titanium and nitrided stainless steel. But, based upon corrosion resistance alone, titanium is by far the best material [9]. The wave machine manufacturer should always be consulted as to the best course of action.

The typical cost for equipment retrofit can range from $15,000 to $35,000 based upon the material choice. Questions are often raised as to whether it is possible to use a wave machine with two different lead-free alloys. Such a solution is not very realistic because in all cases, for each alloy change,

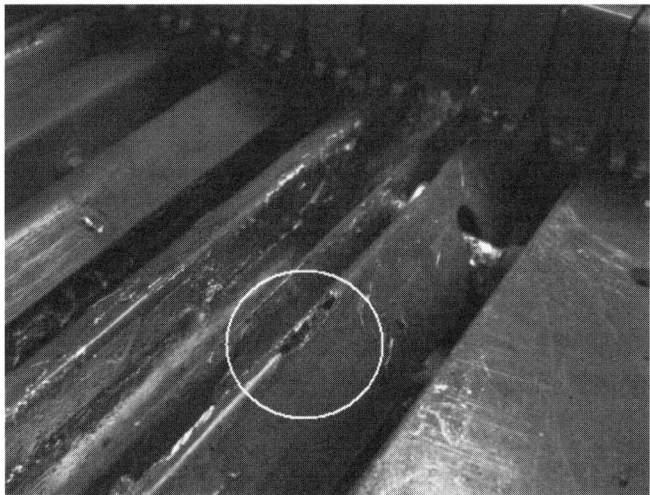

Fig. 4.1. Stainless steel part erosion in a lead-free SnAgCu wave machine solder pot after six months of usage

the entire pot has to be emptied, properly cleaned and refilled. A compromise proposed by certain wave manufacturers is the use of the trolley option with dedicated pots.

It is also important to note that many of the new wave solder pots offer longer contact areas to aid in the wetting of lead-free alloys without necessitating a significant slowdown of the conveyer speed. Companies should weigh the expense of purchasing new equipment versus the lower throughput of older wave soldering equipment [9].

Wave soldering pallets need to be part of the material consideration. With a lead-free process, the pallets will be exposed to longer thermal cycles with slower conveyor speeds and longer contact times in the wave. It is best to check with the wave pallet supplier to verify the material's stability for lead-free wave soldering and today new composite materials are available on the market to address lead-free temperature constraints [10,11].

4.3.2 Solder Pot Analysis

When lead-free wave soldering is set-up in production, it is critical to monitor the alloy composition as well as any accumulation of elements. Copper will have a tendency to increase such as if a lot of OSP finished boards are soldered. The increase of copper has an impact on the alloy melting temperature and on Cu_6Sn_5 intermetallic formation. The problem is that, contrarily to what was observed with SnPb wave alloy, Cu_6Sn_5 intermetallics are found more at the bottom of the pot instead of the top surface as they are denser than SnCu or SnAgCu alloys. Therefore, instead of the intermetallic floating off and being easily removed, the intermetallics are then dispersed through the lead-free alloy in the wave.

Cookson indicated that an average leaching rate of 0.01wt% Cu is observed per 1000 OSP finish boards processed in SAC305 wave alloy at 260°C [12]. The copper accumulation is a function of copper content in the as-received solder bar and board factors, i.e. surface finish type, component type, number of leads and surface area exposed to wave. Contract manufacturing sites purchase solder bar from different suppliers, build boards for different customers with different board complexity. Even within a site, different wave machines may support different customers. Hence, the copper accumulation trend is unique for each plant and even for each wave machine within a site.

It is typically recommended to keep the copper between 0.5 and 0.9wt% for SnAgCu alloys and in any case, if copper becomes higher than 1.0wt%, the liquidus temperature may increase and the alloy will become sluggish causing more solder bridges. At 1.5-1.9wt% copper, precipitation in the pot would occur and may potentially damage the equipment.

Monitoring of elements like lead, (or even iron) is needed. Typically lead contamination can be caused by either wrong solder bar loaded since tin-lead production is typically in parallel to lead-free production or, lead leached from the component lead plating due to poor or no BOM (Bill of Materials) scrubbing or wrong components used. Lead contamination, in addition to the fact that the EU ROHS legislation requires less than 0.1% Pb weight, will decrease the melting temperature and different alloy compositions may form in the solder. For instance, the addition of lead in SnCu solder will result in SnPb alloy formed and its addition in SnAgCu will result in SnPbAg and SnPb alloy formation. Lead-free tin-based solder always contains a low level of lead in the as-received solder bar typically in the range of 0.02 to 0.05wt% (200-500ppm). Before ordering solder alloy, one needs to ask the alloy manufacturer what their maximum allowed lead content is in the raw material.

Iron increase in the solder alloy may also be an issue. In this case, iron will come from the equipment materials that are eroded by the lead-free alloy and above 0.02wt% Fe in the joint can cause it to look gritty [12,13]. For lead and iron accumulation, it has to be understood that it is difficult to remove and the only real solution is to change the lead-free alloy.

To monitor the two elements (Pb and Cu) along with other materials such as cadmium, gold, iron, zinc, aluminium, bismuth and arsenic which are listed in IPC standards such as J-STD-002 [14], the frequency of the sampling analysis should be defined by the use of the equipment and the volume of production. During equipment introduction, one analysis per month should be reasonable and after 6 to 12 months of use, the frequency could be optimised. The decision to add Sn3.0Ag solder bar to balance the copper increase of the SnAgCu solder pot should be made based on laboratory analysis results. Solder analysis can be performed by an external private analytical lab but this service is also often available by the solder alloy manufacturer.

It is also safe to monitor on site the lead contamination in the wave or mini-pot baths with internal lead measurement tools. If the lead levels exceed the European Union RoHS allowable level ($\geq$ 0.1wt% Pb), production needs to be stopped immediately. There is no known method to remove lead from the contaminated pot. There may be a need to drain all the solder out from the contaminated wave pot or solder fountain and replace with fresh solder, which would make this exercise expensive.

4.3.3 Material Segregation

As the transition to lead-free will not be in a single step, assemblers will have to produce during many years, mixed processes associating production with tin-lead and lead-free processes. There will need to be evidence and actions developed during this period on how to avoid cross-contamination. For solder wave, alloy manufacturers can produce bar with different shapes and with additional markings on the individual bar and on the packaging. But the safest solution will be to set-up "Poke yoke" systems to avoid the situation where human error can cause major issues such as during solder bar feeding operations. Solder bars should be stored in separate dedicated areas. Tooling such as spatulas needs to be dedicated and well marked for the lead-free wave machine so that no tin-lead solder will be transferred to the pot for lead-free and vice versa with this including the containers for dross recovery.

Also care has to be taken with handling of tooling and parts close to the lead-free wave solder, as already mentioned, the lower density of the lead-free alloys means that components or tooling that are dropped by error into the solder pot will no longer float to the top as they would for tin-lead.

4.4 Process Recommendations

4.4.1 Process Parameters/Factors

For lead-free soldering, the standard parameters used for tin-lead soldering need to be adapted. Soldering depends on the optimal combination of flux, temperature and alloy and for the wave process the important factors are broken down into:

Solder alloy choice
Flux choice
Fluxing method
Preheat temperature and type
Conveyor speed
Wave shape
Soldering atmosphere
Solder pot temperature
Wave machine
Fluxer equipment

For double sided boards with through hole components the typical process parameters for a no-clean VOC (Volatile Organic Compound)-free flux with SnAgCu solder are shown in Table 4.2.

Table 4.2. Typical process window parameters for lead-free wave soldering with VOC-free flux

Preheat temperature (measured on topside of the board)	110-130°C
Wave pot temperature	265-275°C
Atmosphere (over solder pot area)	Nitrogen
Contact time on chip wave	1-2 seconds
Contact time on laminar wave	3-6 seconds

4.4.1.1 Flux Application

For flux application, development has focussed on spray fluxers that have become a standard. The advantage of the spray fluxer is their enclosed shape and the application of fresh flux on the board with more reproducible quantity (1).

For SnPb wave soldering, any kind of flux can be used from alcohol to VOC-free. For lead-free soldering, in a no-clean process, a water based VOC-free flux is generally recommended, as it is more active and generally more stable for the increased soldering temperature. In addition, the use of VOC-free offers a completely green solution for the wave soldering process. Although spray fluxers are recommended with VOC-free flux, care has to be taken when selecting a flux with the compatibility of the spray fluxer material that can be attacked by the corrosive nature of the flux [15]. While tin-lead wave soldering can be processed with no-clean flux with 1.5% wt solid content, lead-free wave soldering requires flux up to 4.5% wt solid content to reach reasonable results. In all cases, one has to work closely with the flux manufacturer to select a suitable flux product. The required flux quantity may be higher as lead-free solder has higher surface tension so does not wet surfaces so readily.

In some cases, no-clean flux residues tend to create a thick flux skin that can affect pin probe testing. Poor flux application quantity can lead to poor solder wetting while excessive flux will create solder balling and leave flux residues which can be harmful. For some applications, water-soluble fluxes will still be required. To monitor flux penetration into the through hole board, an absorbent paper like thermal fax paper can be placed on the topside of the bare board and passed through the spray fluxing system which applies flux from the bottom side of the board.

4.4.1.2 Preheat

The primary role of the preheat is to prepare for the wave soldering. The preheat has a role in activation of the flux and evaporation of its volatiles (alcohol or water). The preheat stage has the role to minimize the thermal shock before the wave soldering operation. Generally, it is safer for components to keep the thermal gradient lower than a 100-120°C range between the preheat and solder pot temperature to avoid thermal shock. It is well known that the ramp up during the preheat stage should be less than 2°C/s. The use of VOC-free fluxes necessitates the use of higher preheat temperature often up to 130°C on the assembly top side and as the amount of energy required to evaporate water increases, this implies a longer preheat stage compared to alcohol based fluxes [13]. To reach the required higher temperatures and minimize the thermal gradient, a longer preheating section is needed.

Although a longer preheat will activate the flux, evaporate solvents and/ or water and minimize thermal gradients, an excessive preheat can be detrimental to the flux and reduce its activity during wave soldering. In addition, the use of too much preheat temperature can induce secondary reflow of topside components. For water based VOC-free fluxes, convection preheat is recommended as it improves the dissipation of the volatiles and provides more uniform heating. But, quartz heaters are beginning to grow in popularity as they offer more flexibility because they react more rapidly. They are therefore well adapted for high mix production environment as they increase line productivity.

4.4.2 Wave Soldering Pot Temperature and Contact Time

The two major differences between lead-free and tin-lead wave soldering are the solder pot temperature and solder contact time. Some recommend the increase of the laminar wave to improve the wetting and reduce the distance between chip and laminar waves to minimize the temperature drop between the two waves [10,11]. As well explained by Shea et al. [10], after passing through the chip wave, solder joints cool and begin to solidify. When they reach the smooth main wave, they are reheated and re-melted. Once the solder again becomes molten, its wetting forces can act on the pin and the barrel to fill the hole. The shorter distance between nozzles gives the lead-free solder less time to cool, and requires less energy to re-melt the solder in the barrels, thus providing more contact time on the smooth wave for wetting.

The net result of closer wave nozzles is better hole fill [10]. The time the flux is exposed to the wave temperature from the entrance to the chip to the exit of the Lambda is also decreased so the survival of the flux increases through the second wave. This has been proven both by equipment manufacturers [10] and in contract manufacturer production plants [11]. The pot temperature, for SnAgCu solder, is equal to 265±5°C while it is

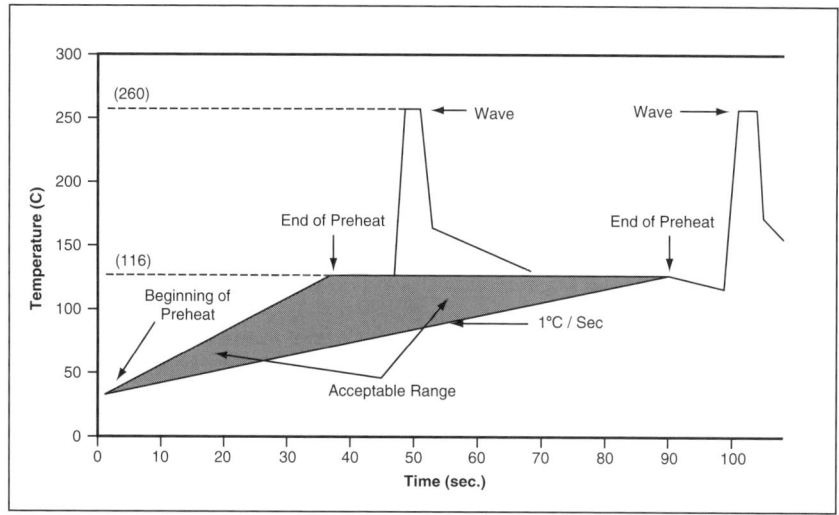

Fig. 4.2. Typical wave thermal profile window for lead-free soldering with VOC-free fluxes. Courtesy Kester [16]

around 10°C higher for SnCu alloy. The correct temperature is really dependant on the board layout and whether a pallet is used or not. But typically, the influence of the process parameters like flux quantity, conveyor speed, dwell time, contact length, component orientation and soldering direction on the quality of the wave soldering are similar for lead-free and tin-lead solders.

Accurate profiling for each new assembly is the key to success and temperature should also be monitored and controlled on the PCB and component topside. Figure 4.2 shows a typical wave profile window recommendation for VOC-free flux [16].

4.4.2.1 Nitrogen Use

Lead-free alloys oxidize more rapidly than Sn37Pb when the solder is in its liquid state due to the increased tin content. Tin oxide (SnO and SnO_2) forms at a higher rate because of the higher processing temperatures, resulting in more oxidation and dross [17]. But a fully inerted machine is not required as the use of nitrogen in the solder pot area, acts like a nitrogen blanket, minimizes exposure of the liquid solder to oxygen and therefore decreasing the amount of dross. In addition, defects like spikes or icicles will not be observed if there is enough flux activity and/or a nitrogen atmosphere.

Without nitrogen, flux residues may be more brown coloured and the oxide formation on the surface of the alloys creates a yellow appearance, which has no effect on the solder joint properties themselves but may cause difficulty for solder joint visual inspection. Just as with tin-lead soldering, an inert atmosphere will improve wetting and open the process window. On the other hand, as also known for the SnPb process, nitrogen inerting increases the amount of solder balls. In addition nitrogen use is expensive.

4.5 Solder Joint Characterization

Volumetric contraction during solidification is a characteristic of high tin lead-free solder alloys and will influence the microstructure of the solder joint. This volumetric contraction is commonly called solder shrinkage. Several commonly observed phenomena such as surface roughness, shrinkage grooves, fillet lifting and hot tearing are attributed to volumetric contraction which is mainly observed in pin through-hole (PTH) joints but can also exist in surface mount technology [18].

4.5.1 Solder Joint Appearance

Lead-free solders have a grainy appearance with a dull texture compared to the typical smooth and shiny appearance of SnPb solder as shown in the IPC-A-610-D standard [19]. Lead free alloys are more likely to have surface roughness (grainy or dull) and greater wetting contact angles. But all other solder fillet criteria are the same [19]. This grainy appearance for lead-free high tin solders is a question for solder joint inspection with criteria/education of operators and inspectors.

The texture variation has been studied by various authors. It is reported from simple experiments on Sn3.0Ag0.5Cu solder with different cooling conditions that a slow cooling (for instance a cooling rate of 0.20°C/sec in the 260-200°C temperature range) leads to a grainy and dull appearance of the solder alloy whereas a rapid cooling (in the range of a cooling rate at 5°C/sec down to 200°C) allows the formation of a smooth, bright and regular surface [20]. This observation applies on all the classically used SnAgCu compositions from Sn3.0Ag0.5Cu [20] to Sn3.8Ag0.7Cu [3]. For SnCuNi type alloys the cooling rate has little or no effect on the solder texture, just as for tin-lead solder. For the Sn3.0-4.0Ag0.5-0.7Cu alloy family, as a consequence of the different intermetallic formation (Cu_6Sn_5, Ag_3Sn and $Sn+Ag_3Sn+Cu_6Sn_5$) their texture has a grainy, even irregular aspect with typical shrinkage grooves, hot tearing or fillet lifting formation. If component leads or the PCB surface finish contains lead, SnPb and even SnPbAg may form.

Classical defects like non-wetting, insufficient solder, solder balls, solder bridges, icicles, which are already known for tin-lead processes are still present or even reinforced with lead-free wave soldering. Others like shrinkage grooves, hot tearing, fillet lifting or pad lifting have become more typical for lead-free soldering.

4.5.2 Solder Balling

From the IPC-A-610D standard, solder balls are spheres of solder that remain after the soldering process. Solder balls that become entrapped/ encapsulated and do not violate minimum electrical clearances are considered as acceptable for IPC Classes 1,2,3 [19].

As already known for tin-lead soldering, nitrogen use increases the amount of solder balls. One major parameter for solder ball formation is the type of board solder mask used. The same case is also for tin-lead soldering, depending on the nature of the solder mask, solder balls may adhere more easily. It is recommended to use a matte type of solder mask. Poor drying of the flux may also create solder balling; similar to any out gassing phenomenon occurring on the PCB.

4.5.3 Solder Bridging

It is recognized that pot temperature has little or no effect on solder bridge formation. The process parameters that more influence their formation are first the wave contact time followed by the preheat temperature. Flux amount is the secondary parameter [21].

4.5.4 Voids

Voids are created from various sources such as flux, unclean surfaces, temperature and moisture. The surface tension is much higher for Pb-free SnAgCu alloys than SnPb so the SnAgCu solder will tend to hold in the voids longer. Regardless of pin/hole area ratios, if a component pin is too close to the board barrel wall, voids will connect to both the pin and barrel surfaces. Rough surfaces have more surface area and are more difficult for voids to break free from. Other factors contributing to voids not escaping the wave soldered joint include the fact that the SnAgCu alloy pot temperature is closer to its solidification temperature (217°C) than SnPb (183°C), thereby there is less time for a void to escape. Typically, the solder pot temperature is around 265°C for SnAgCu alloys whilst it is around 255°C

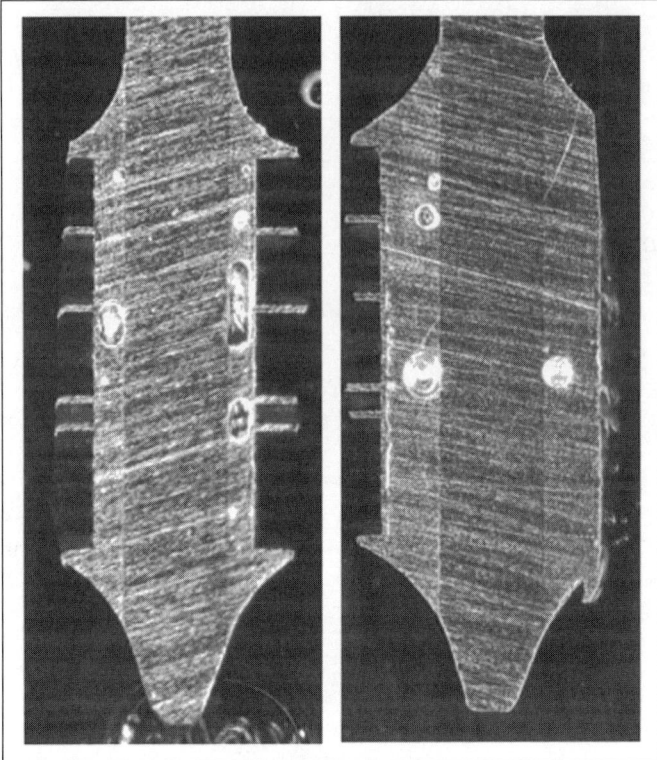

Fig. 4.3. Cross-sectional analysis for lead-free SnAgCu soldered PTH solder joints showing larger holes (0.047 inch diameter) with a low pin to hole area ratio of 0.65 (right hand picture) has less voids than small holes (0.035 inch diameter) with a high pin to hole area ratio of 0.36 (left hand picture) for the same pin cross-section (0.025 inch square pin) [20]

for SnPb. All parameters being equal, larger holes with low pin to hole area ratios have less voids as shown in Figure 4.3.

If the leads are used for high power resistance or temperature dissipation, voids could be an issue. For other applications, voids in PTH solder joints are not recognized to cause any issue [7].

4.5.5 Fillet Lifting

Fillet lifting is caused by the large Z-axis expansion of the board material at soldering temperatures where the solder fillet lifts from the copper pad. Since the soldering temperature with lead-free solder is higher than with tin-lead solder this expansion is more important. An illustration of fillet lifting from cross-sectioning is shown in Figure 4.4 [22]. Sometimes, fillet lifting can occur at the pin–solder interfaces. It has been observed on PTH waved solder joints and is rarely observed with the reflow process. Separation occurs between the intermetallic and solder and it is reported that the fillet lift stops at the knee on the land side. Again, it is more frequently observed with temperature pasty range alloys where the alloy solidifies gradually instead of rapidly like eutectic alloys [3]. This phenomenon is not visible with the naked eye and even with a low magnification power microscope.

Chapter 5 of the IPC-A-610D standard [19] proposes acceptance criteria for fillet lifting for Classes 1, 2 and 3 where there is separation of the bottom of the solder near the land on the primary side of the plated through hole connection. There is a defect for all classes where the fillet lift damages the land attachment or for Class 3 only where there is separation of the bottom of the solder near the land on the secondary side of the plated through hole connection.

In addition to the internal constraints in the alloy during the pasty range for solidification, the movement of the PCB and that of the components introduce other mechanical constraints and increase the stress in the solder joint. Stresses in the solder joint are due to the mismatch of the different CTEs in the used materials. Depending on if the weakest point location is on the pad to dielectric interface, pad lifting may form, but if the strength on the PCB is greater than that of the solder, shrinkage or even fillet lifting may form [18, 23].

Fig. 4.4. Fillet Lifting [22]

It has been explained that preventing bismuth and lead contamination will minimize fillet lifting [22, 24]. However, this prevention does not appear to be sufficient. Fillet lifting is an intrinsic phenomenon of lead-free alloys similar to hot tearing. In fact, before the last part of the liquid alloy solidifies, the main part which has solidified can loose contact with the pad. As explained by Lee [3], the main cause of fillet lifting is attributed to CTE mismatch: during cooling from liquid to solid stage, with solder producing a tearing shearing force in the x-y direction due to its higher CTE than the PCB, while the PCB itself generates a tearing tension in the z-direction, due this time to its higher z-CTE than solder. As a consequence, fillet lifting is formed. It has been shown that the most critical factor for fillet lifting formation is the temperature difference during solidification [25]. The thermal phenomenon has been demonstrated by NIST/ University of Greenwich research [25].

4.5.6 Pad Lifting

The solder pad has a wedge shaped form, because the thermal expansion of the copper barrel on which the solder pad is connected, is much smaller than the base material under the copper pad as shown in Figure 4.5. From internal experiments on classical FR4 PCB materials, it has been shown that pad lifting was systematically observed on square pads. Today, one

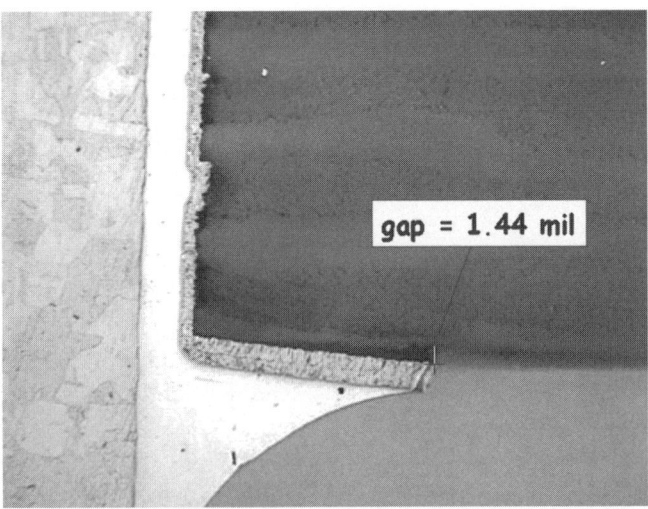

Fig. 4.5. Pad lifting on a lead-free coated component on a square pad using a standard FR4 PCB. The vertical pad lift gap was 1.44mils. Wave solder alloy used was Sn3Ag0.5Cu. 100X magnification [20]

recommendation is not to use square pads on the bottom side of the board. Again, use of materials with compatible expansion coefficients may help in minimising pad lifting formation.

Some have reported that pad lifting occurs more often with SnCu and SnCuNi alloys compared with SnAgCu perhaps because of the increased liquidus temperature of SnCu/SnCuNi compared with increased constraints during the solidification stage [26].

From IPC-A-610D standard, pad lifting is reported as a defect in IPC Classes 1,2,3 when separation between conductor or land and laminate surface is greater than one pad thickness. Any lifting of a land if there is a via in the land is considered a defect for IPC Class 3 [19].

4.5.7 Hot Tearing and Shrinkage Grooves

During solidification, the solder will shrink by about 4% by volume which can form hot tearing/micro-grooves in the solder joint. The hot tearing phenomenon is intrinsic for higher tin-containing SnAgCu alloys and experiments made on solder bar fused and cooled down have shown that they will systematically form under standard cooling conditions. In fact, due to the different stages of solidification, the tear will appear in the warmest area of the solder which is the last area to solidify. Experiments have shown that hot tearing can be avoiding when the SnAgCu alloy is cooled down to 200°C at a rapid cool down rate (at least 5°C/s) leading to a shiny appearance [20].

The CTE mismatch is important because for either the PCB material, component size and height, component material (such as ceramic) or a combination of the above, the tear can expand to shrinkage grooves [27]. Implementing forced cooling would help reduce shrinkage groove formation but the cooling action should be set just after the wave solder within 2 to 3 seconds after the board has left the wave solder [11]. One can easily understand that it is difficult or even impossible to set this up on a wave machine as the forced air would cool down the wave pot itself [20, 23].

Hot tearing and shrinkage grooves are generally found at the surface of the solder joint, as shown in Figure 4.6. Hot tearing is commonly observed in lead-free PTH solder joints, and also on some BGA lead-free solder joints. For PTH solder joints, they can be observed in solder joints with components coated with either lead-free or tin-lead material, in filled holes without components or via holes, or on the top or bottom side of the assemblies.

Shrinkage groove formation is directly linked to the solder volume and will appear mainly in the solder bulk near large copper annular rings where the solidification takes longer. During SnAgCu cooling, 3 phases are formed during the solidification: β-Sn, Ag_3Sn and Cu_6Sn_5 which have different

Fig. 4.6. Shrinkage groove

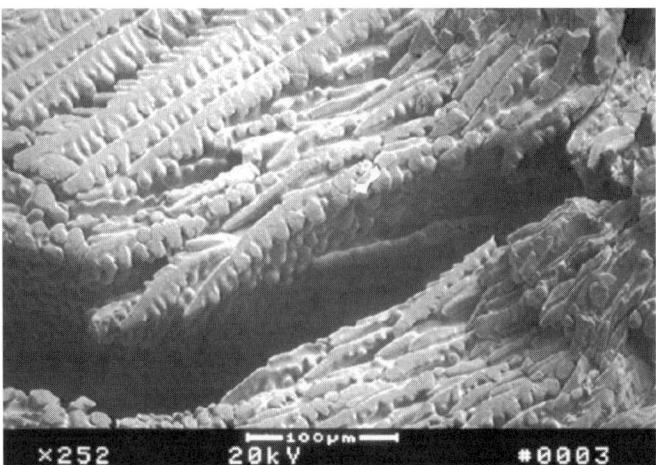

Fig. 4.7. SEM view of shrinkage grooves in a lead-free SnAgCu wave soldered joint

thermal or physical properties. The hot tearing appears parallel to the Ag_3Sn needles and surrounded by β-Sn phase (Figure 4.7).

A fairly recent publication gave an explanation for the shrinkage groove formation mechanism, which is shown schematically in Figure 4.8 [28]. In this area, Sweatman et al. [29] also carried out studies on solidification of Sn0.7Cu alloy which is the eutectic composition and explained that shrinkage grooves formed because of the higher tin rich content in the alloy and with the grooves appearing with β-Sn dendrites that increased during cooling

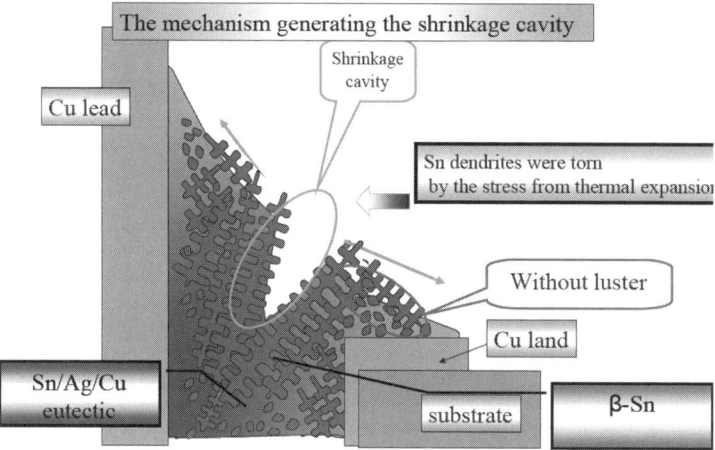

Fig. 4.8. Shrinkage groove formation mechanism. Courtesy Senju Metal Industry Co. Ltd [28]

and not because of the eutectic composition of the alloy. The basic Sn0.7Cu alloy did not behave like the eutectic that it was supposed to be, contrary to the eutectic Sn37Pb composition [29]. Shrinkage grooves are also observed in Sn3.5Ag which has the eutectic composition at 221°C and confirms that these shrinkage grooves are not only linked to a non eutectic composition but principally because of the higher tin content in the alloy [29, 30].

From the IPC-A-610D standard [19], hot tearing/shrinkage grooves are acceptable for Classes 1,2,3 for connections made with lead free alloys where the bottom of the tear is visible or the tear or shrink hole does not contact the lead, land or barrel wall. They are considered as defects for Classes 1,2,3 where there are shrinkage grooves or hot tears in connections made with SnPb solder alloys or for connections made with lead free alloys where the bottom of the shrinkage groove or hot tear is not visible or the tear or shrinkage groove contacts the lead or PCB land. Comparing SnCuNi with SnAgCu alloy the SnCuNi solder is not prone to shrinkage groove formation. On the other hand, more pad lifting could be observed with SnCuNi alloy compared to SnAgCu wave soldered boards.

4.5.8 Through Hole Filling

Through hole fill requirements have not changed from IPC-A-610C to IPC-A-610D [19]. For tin-lead or lead-free Class 2 assemblies, the acceptance criteria is 75% hole fill but 50% is accepted when barrels are connected to power or ground planes [19].

Poor through hole filling is a typical problem for lead-free wave soldering. Obviously, longer solder contact time and/or higher solder temperature result in better through-hole filling. Taking OSP PCB surface finish boards, which are known to have poor hole fill ability, design of experiment studies have shown that many factors affect hole fill such as pot temperature, flux amount, preheat temperature, conveyor speed or even pin protrusion, but one of the main factors that influence hole fill is the flux type. Along with flux type, it has been demonstrated that preheat temperature (130°C) affects the process the most while the impact of the other factors is about equal [21].

Higher solder temperatures also will benefit the wetting of the joints but the combination of longer contact times and higher solder temperature is an issue since they can lead to several defects like blow holes, solder excess, copper dissolution/leaching (which is recognized as being an increased concern with SnAgCu alloys compared with SnPb) or even secondary reflow, solder wicking, or thermal damage to components or the board [31]. In some cases, the surface mount adhesive for bottom side wave soldered surface mount components cannot withstand the higher lead-free wave soldering temperatures and the bottom side components may drop into the wave solder.

4.6 Design Considerations

Many areas need to be considered during design for lead-free wave soldering which are discussed in the following section.

4.6.1 PCB Surface Finish Comparisons Related to Lead-Free Wave Soldering

Table 4.3 gives the typical PCB finishes available today on the market for lead-free soldering which are discussed in more detail in the PCB surface

Table 4.3. Lead-free PCB surface finishes

Lead-free Hot Air Solder Level (HASL)
Organic Solderability Preservative (OSP): standard or high temperature rated
Electroless Nickel Immersion Gold (ENIG)
Electrolytic NiAu
Electroless NiPdAu
Immersion Silver
Immersion Tin
OSP with Selective ENIG
Electroless Palladium

finish chapter of this book. There can be solderability issues for the different board surface finishes as discussed below.

4.6.1.1 Electroless Nickel Immersion Gold (ENIG)

Gold makes an excellent surface finish since it dissolves rapidly in lead-free solder and does not tarnish or oxidize. A porous gold layer or the black pad phenomenon will affect the solderability of the layer that can be observed even after a first reflow soldering operation. In addition, NiSn intermetallics are formed with the nickel layer compared with SnCu intermetallic with the other major PCB finishes (OSP, Ag, Sn, Lead-free HASL) and are recognized to be more brittle than the SnCu intermetallic.

4.6.1.2 Immersion Tin

For this surface finish, SnCu intermetallic layers are formed between the pure tin layer and the copper pads in the as-received board. If the intermetallic reaches the coating surface, it oxidizes rapidly resulting in poor solderability of the solder joint area. Surface morphology studies show that the immersion Sn can have a porous and grainy structure. A tin thickness equal to 1.2μm is typically required to allow good soldering for lead-free processes when reflow soldering is done prior to the wave operation. Immersion tin finish is not typically recommended in lead-free processes where double sided SMT reflow is followed by wave soldering as the tin layer may have all been consumed before the wave operation leading to solderability issues.

4.6.1.3 Immersion Silver

This surface finish has a relatively thin deposit. The pins of the test inspection probe tools easily damage this very thin layer. Some yellowing during the processing may occur depending on the amount of oxidation in air. Darker tarnish layers of silver sulfide Ag_2S caused by atmosphere or contact contamination may affect the solderability.

4.6.1.4 Organic Solderability Preservatives (OSP)

OSPs must keep the copper surface solderable prior to and during the soldering process. Over time, unprotected Cu surfaces are oxidized, leading to the formation of Cu_2O, and later, a mixture of Cu_2O and CuO. OSP finishes are usually recommended for lead-free wave soldering but not for thicker boards greater than 93 mils thick where hole fill is generally poorer and test points are not systematically covered with solder. For this last point, where test probes make poorer contact, Intel found that that single sided SMT followed by wave solder assembly with OSP resulted in 2-4% unsoldered test points on the solder side of board [32].

The constraints of the OSP finish are also linked to handling which is not easy to control in PCB assembly as there needs to be storage time control and a controlled time between each process step which is difficult to do. This makes the OSP finish to be not as adaptable for a low volume or highly mixed production environment. If OSP finish is requested, then the high temperature rated OSP coating needs to be ordered and actually provided. In this case, as the process window will be narrower, it will be more important as mentioned by Holder et al. [33] to keep the time for wave processing after SMT assembly to a minimum, which will be no longer than 48 hours from first thermal exposure and if possible little or no holding time after SMT reflow before lead-free wave soldering is conducted.

4.6.2 Temperature Related Issues for the Board and Board Surface Finish

These include areas such as the higher soldering temperature contributing to a number of potential failures. Delamination, separation of layers of the base material and/or between the base material and copper and board warpage are typical defects due to reduced board quality in combination with the higher soldering temperatures. Care has to be taken with the PCB baking of OSP or immersion tin finish boards prior to assembly. In the first case, the organic protection starts to be removed during the baking and soldering issues may occur if boards are not quickly processed after baking. For the immersion tin finish, storage at temperature increases the SnCu intermetallic thickness formed and when the total tin layer is consumed, good soldering cannot occur. For lead-free PCB surface finish alternatives, some companies such as Intel recommend immersion silver PCB finish [32] especially for thicker boards for lead-free wave soldering.

4.6.3 PCB Design Rules

Existing design rules established for SnPb should be continued for lead-free assembly. As hole fill is more of an issue with the lead-free wave process, designers strongly need to keep in mind the use of thermal planes only if mandatory for the product application. If thermal ground planes are used, their designs should be optimised to improve hole fill.

A small pin to hole area ratio (with a typical Pin to Hole area ratio of 0.35) will allow to minimize void formation in the PTH barrel but increasing solder volume is recognized to induce more shrinkage groove formation. A compromise has to be found with small pin to hole ratios and minimum copper pad diameters. Square through holes copper pads have been demonstrated to be more sensitive to pad lifting formation. It is recommended to avoid their use on the bottom side of the board in any new design for the lead-free soldering process.

4.6.4 PTH Solder Joint Reliability

Poor hole fill is a major soldering issue with lead-free or tin-lead wave soldering and excess contact times to solve this problem for wave soldering or wave soldering mini-pot rework repair can affect the reliability of the final product as there are excessive thermal constraints on the PCB or the components and copper dissolution of the barrel knee and pad lifting may occur. Moreover, all the additional operations performed to compensate poor hole fill are time consuming and may generate more scrap.

A recent study presented by Ferrer et al. has shown that IPC-A-610D hole fill criteria are conservative [34]. Obviously, as an end-user, it is recommended that process engineers continue to improve their processes with the objective to get 100% hole fill. But, in some cases, when all process options have been set up and optimized, 75% hole fill without thermal planes and 50% hole fill with thermal planes for Class 2 are not realistic and insisting to obtain this target may generate more problems. So, engineers and end users need to work together and finally allow lower hole filling in some cases. In the study, HP showed that for Class 2 and 3 IPC products, for a 130 mil PCB thickness, 36% SnAgCu hole fill was sufficient and reliable and that for a 97 mil PCB thickness, 48% SnAgCu hole fill was satisfactory based on reliability testing [34].

Lead addition in SnAgCu alloy does not appear to affect pull testing results when lead contamination is kept below 1% wt [35] but above 0.1wt% lead does not conform to the EU ROHS legislation limit. Some sources indicate that reliability of alloys containing silver such as SnAgCu offer better reliability compared to Sn0.7Cu or Sn0.7CuNi alloys because of the presence of Ag_3Sn intermetallic plates that can act as a barrier to crack propagation which arrests or redirects crack propagation when oriented transverse to the large grain boundaries in the solder [36]. However, from the final report of the no-lead solder project of JCAA/JG-PP consortium, no difference was detected between Sn3.9Ag0.6Cu, Sn0.7CuNi and Sn37Pb wave soldered joints compared under high stress conditions (thermal shock −55°C to +125°C, 15min dwell) [37].

It was demonstrated by Ueshima et al.[28] that more and larger shrinkage grooves are observed in Sn3.0Ag0.5Cu compared to Sn4.0AgxCu (0.5 <x< 1.2) due to the fact that the β-Sn dendrite decreases as Ag and Cu content increases. However, from different studies, it is important to mention that thermal cycle cracking or solder joint failures were not found to be associated with the shrinkage grooves observed [37,38,39].

It is also claimed that fillet lifting is also a benign phenomenon and that only one type of lifting, pad lifting, is malignant. In the latter, peeling of the pad from the substrate could destroy the connection, an issue which Japan manufacturers are addressing in terms of redesign of pads and solder mask patterns [23,27]. But, this needs to be confirmed with further studies mainly on high end applications where the thermal excursion of the product may be high during its life.

Conclusions

Wave soldering is a common soldering technique used to solder through holes components and some SMD components on the PCB. With the traditional tin-lead (Sn37Pb) alloy being steadily replaced by lead-free solder, today, the electronic industry is recommending Sn3.0Ag0.5Cu as the lead-free alloy alternative. Due to cost issues, SN100C (Sn0.6Cu0.05Ni) and similar alloys are becoming interesting alternatives.

The wave solder machine needs to be adapted to lead-free soldering to avoid erosion of all its machine parts that are in contact with the molten alloy which should be considered to be retrofitted. During a transition that can range from years to decades, the electronics manufacturing shop floor will need to live with leaded and lead-free wave solder machines and materials segregation will be one of the most challenging items where cross-contamination will need to be avoided.

Frequency of bath alloy analysis will need to be reinforced with careful analysis of the elements such as lead, copper and iron. From the process point of view, the know-how gained with tin-lead soldering can be transposed to lead-free solder processes but typically the lead-free wave process window will be narrowed and the risk of damage of components and/or PCB increased. Higher solids content no-clean fluxes will give good results for lead-free wave soldering but there would be an issue with test pin probeability results. If assemblers are not used to VOC-free fluxes, they will need education as they are typically preferred for lead-free soldering as they offer a complete green solution. With these water based fluxes, increased preheat will be more critical to insure water removal.

Lead-free solder joints have a grainy appearance and some anomalies/defects may increase with lead-free soldering such as voiding , hot tearing and shrinkage grooves, fillet lifting, pad lifting and hole fill. Although the European Union ROHS July 1st 2006 legislation date has passed, studies have still to be performed upon final product application as the effect of these potential anomalies/defects on solder joint long term reliability is not well known.

Acknowledgements

The principal author, Christiane Faure, dedicates this chapter to Marie and Nicolas. The authors wish to thank numerous colleagues from Solectron past and present for their continuous technical guidance and encouragement over the years and for this work especially Jean-François Couderc, Gilbert Zanon, Dennis Willie and Mark Elkins.

References

1. Friedrich J (2003) Practical experience in lead-free wave soldering, ERSA GmbH
2. Moon K-W, Boettinger WJ, Kattner UR, Biancaniello FS, Handwerker CA (2000) Experimental and thermodynamics assessment of SnAgCu alloys, J. Electron. Mater., Vol. 29, p1122-1236
3. Lee N-C, Reflow soldering processes and troubleshooting SMT, BGA, CSP and Flip-Chip Technologies, Newnes, 12-253
4. Sweatman K (2005) Another chance for tin-copper as a lead-free solder. APEX Special Issue, p 4-7
5. Nimmo K (2003) Second European Lead-free Soldering Technology Roadmap, Soldertec at Tin Technology Ltd, p11-12
6. Seelig K (1998) A comparison of leading lead-free alloys, AIM information,Volume & Issue1, http://www.aimsolder.com/technical_articles.cfm#17
7. Alpha Metals, Vaculoy SACX0307, Lead-free solder wave alloy technical bulletin, US Patent 4929423
8. Seelig K (2005) Considerations for lead-free wave soldering AIM information, http://www.aimsolder.com/technical_articles
9. Morris J, O'Keefe M J (2004) Equipment impact of lead-free wave soldering, Appliance Magazine
10. Shea C, Barton B, Belmonte J, Kirby K (2005) Practical lead-free implementation, http://www.speedlinetech.com/docs/Practical-LF-Implementation.pdf
11. Couderc J-F (2005) Solectron France lead-free wave soldering internal report
12. Alpha Vaculoy SAC300, 305, 400, 405 Ultra low lead free wave solder alloy, Cookson Electronics, Alpha Technical bulletin
13. Barbini D, Wang P (2005) Implementing lead-free wave soldering iNEMI, Printed Circuit Design and Manufacture.
14. J-STD-002B Standard (2003) Solderability Tests for Component Leads, Terminations, Lugs, Terminals and Wires
15. Wang X, Poon C (2003) Benchmarking of lead-free no-clean fluxes soldering performance on OSP boards, SMTAI Conf Proceedings, Rosemont, USA
16. Kester VOC free flux technical bulletin
17. Gyemant T (2004) Lead-free wave soldering: a cost-effective alternative, SMT Articles

18. Wable GS, Chada S, Neal B, Fournelle RA (2005) Solidification shrinkage defects in electronic solders, Journal Of Materials, Vol. 57, no. 6, p38-42
19. IPC-A-610D Standard (2005) Acceptability of Electronics Assemblies
20. Faure C (2005) Lead-free Assembly: Process Considerations, IPC /Soldertec conference, Barcelona, Spain
21. Zetech (2005) Analyzing lead-free soldering defects in wave soldering using Taguchi methods, Dataweek–Electronics & Communications Technology
22. Bath J, Hueste G (2001) Lead-free Sn3.5Ag and Sn0.7Cu Wave Solder Evaluation with VOC-Free No-Clean and Water Soluble Fluxes, Nepcon East Conference, Boston, USA
23. Diepstraten G (2006) Matte or dull joints are normal and should be considered just an effect, Circuit Assembly magazine
24. BE95 (1994) IDEALS, BRITE-EURAM program, Synthesis report
25. Handwerker C (2002) NIST Research in Lead-Free Solders: Properties, Processing, Reliability, UC Smart
26. Havia E, Montonen H, Bernhardt E, Alatalo M (2005) Comparing SAC and SnCuNi solders in lead-free wave soldering process, NEXT Symposium, Finland.
27. Schouten G, Pad lifting, fillet lifting and fillet tearing issues in lead-free soldering, Vitronics-Soltec Technical White Paper
28. Ueshima M, Nodera M, Tajika T (2006) Mechanism of shrinkage cavities and method for restricting them in SnAgCu alloy system IPC-Soldertec conference on lead-free electronics, Malmö, Sweden
29. Sweatman K, Nishimura T, Jost J (2006) Solidification behaviour of the Ni-modified Sn0.7Cu eutectic, IPC-Soldertec conference on lead-free electronics, Malmö, Sweden
30. Handwerker CA, Noctor D, Whitten G (2000) Current assessment of the reliability of lead-free solders
31. Diepstraten G, The Range and Causes of Lead-Free Soldering Defects, Vitronics-Soltec Technical White Paper
32. Long G (2006) Lead-free surface finish overview, Intel Lead Free Symposium Sponsored by IPC, Scottsdale, Arizona, USA
33. Holder H, Billaut F (2006) OSP as a Pb-free surface finish at HP, Intel Lead Free Symposium Sponsored by IPC, Scottsdale, Arizona, USA
34. Ferrer E, Benedetto E, Freedman G, Billaut F, Holder H, Gonzalez D (2006) Reliability of Partially Filled SAC305 Through Hole Joints, IPC Printed Circuits Expo, APEX and the Designers Summit
35. Biocca P (2005) Lead-free reliability–building it right the first time, Kester, http://www.emsnow.com
36. Boone L, Campbell GP, Palacio LC (2005) Lead-free solder alloy selection: reliability is a key, SMT
37. Hillman D, Wilcox R (2006) JCAA/JG-PP No-Lead Solder Project:-55°C to +125°C Thermal Cycle Testing Final Report Revision B, Rockwell Collins Advanced Manufacturing Technology Group
38. NCMS Report 0401RE96 Pb-free Solder Project Final Report, (1997) National Center for Manufacturing Sciences
39. Faure C, Elkins M, Willie D, Zhou X, Lamb M, Chatterji I, Mainwaring S, Bath J Lead-free Pin Through Hole Reliability (2006), Solectron Internal Report

Chapter 5: Lead-Free Rework

Jasbir Bath, Solectron Corporation, Milpitas, California

5.1 Introduction

With tin-lead soldering, there is a long history of soldering experience from hand soldering to wave soldering to surface mount technology. The development of lead-free solder manufacturing experience has been a relatively recent occurrence. For lead-free rework the developments that have occurred have not been reviewed in a comprehensive manner.

This chapter will overview rework of lead-free solder for surface mount and wave soldering. Studies evaluating the temperatures that will likely occur on boards and components during lead-free rework will be reviewed. This chapter will aim to give a status on the development work programs in the areas mentioned and indicate where there is a need for future work.

This chapter is divided into four sections:

1. Lead-free SMT hand soldering rework
2. Lead-free SMT BGA/CSP rework
3. Lead-free PTH hand soldering rework
4. Lead-free PTH mini-pot soldering rework

5.2 Lead-Free Hand Soldering SMT Rework

This section is broken down into the following areas:

- Alloy choices
- Solderability and soldering temperatures
- Solder iron tip temperatures
- Solder iron tip life
- Visual inspection of reworked SMT solder joints
- Hand Solder Reliability

- Hand Solder Training
- Conclusions
- Future work

5.2.1 Alloy Choices

The choice of lead-free solder rework alloy is usually dependent on reflow and wave alloy selection. As the assembly materials for surface mount are Sn3Ag0.5Cu and for wave soldering are Sn3Ag0.5Cu or Sn0.7Cu or Sn3.5Ag, the hand soldering rework material could be Sn3Ag0.5Cu or Sn0.7Cu or Sn3.5Ag with all three lead-free alloys having melting points in the similar temperature range (217°C to 227°C).

An experiment was conducted to determine one alloy to choose for rework from three lead-free wire materials (SnAgCu, SnAg, SnCu) with no-clean liquid rework flux on SnAgCu surface mount soldered joints [1]. Tin-lead wire was used to rework SnPb soldered boards as a control.

The same hand soldering rework equipment was used for lead-free and tin-lead rework with a change in soldering tip to avoid lead contamination during lead-free hand soldering. The temperature settings on the hand soldering equipment were raised slightly for lead-free compared to tin-lead rework. This was because the melting temperature of the lead-free solder wire was around 35°C above that for tin-lead (217°C for SnAgCu, 221°C for Sn3.5Ag, 227°C for Sn0.7Cu compared with 183°C for tin-lead). Visual inspection of all the lead-free rework soldered joints were acceptable indicating any of the three lead-free alloys could be used.

Tsang [2] showed that when lead-free Sn4Ag and Sn3.8Ag0.7Cu solder rework wire was evaluated for lead-free solder rework against one another either alloy could be used. Work from the IDEALS Lead-free European project showed Sn3.5Ag wire to be effective for reworking lead-free SnAgCu surface mount boards with the joints looking less 'grainy' than when rework was conducted with SnAgCu solder wire [3].

As the SnAg rework wire had a history of successful use in the industry it would typically be recommended as the rework material of choice to account for the range of lead-free solder alloy joints to rework including SnAgCu, SnCu and SnAg.

5.2.2 Solderability and Soldering Temperatures

In order to form a good solder joint, the solder alloy used for the interconnection must have good solderability. Poor wetting for hand soldering rework can be attributed to insufficient temperature and contact time during soldering and insufficient solderability of the surfaces to be soldered.

Tin-lead solder Sn37Pb solder melts at 183°C. For tin-lead reflow soldering we typically have the solder joint reaching 205°C to 220°C peak temperature with 45 to 75 seconds over the melting point to ensure a good solder joint formed. Excessive temperatures and times can cause excessive intermetallic compounds to form, which can be a detriment to reliability, and the increased temperature can also bake out the soldering fluxes used in the solder paste leading to soldering issues. For tin-lead wave soldering the typical solder pot temperatures used are 250°C to 260°C with 3 to 4 seconds solder contact time in the wave. For tin-lead hand soldering rework it is typical to use a soldering iron tip temperature at 333°C (650°F) with a 2 to 5 second soldering iron contact time at the solder joint.

Lead-free solder Sn3Ag0.5Cu solder melts at 217°C. For lead-free reflow soldering we typically have the solder joint reaching 235°C to 250°C peak temperature with 45 to 75 seconds over the melting point to ensure a good solder joint formed. Again excessive temperatures and times can cause excessive intermetallic compounds formed which can be a detriment to reliability and these increased temperatures can also bake out the soldering fluxes used in the solder paste leading to soldering issues. For lead-free SnAgCu wave soldering typical solder pot temperatures used are 260°C to 270°C with 3 to 4 seconds solder contact time in the wave. For lead-free SnAg hand soldering rework it is typical to use a soldering iron tip temperature at 371°C (700°F) with a 2 to 5 second soldering iron contact time at the solder joint.

5.2.3 Solder Iron Tip Temperatures

The two factors affecting hand soldering quality are operator skill and soldering iron efficiency. During rework of solder joints the solder iron tip temperature should remain fairly constant. Soldering irons generally do not recover lost heat fast enough with rework operators often using higher set temperatures (380°C to 440°C) which can damage the board and components.

When solder tip temperatures are not high enough, or when flux activation is insufficient, poor wetting can occur. Excessive solder tip temperatures may result in dewetting and thermal damage to boards and components. Using the correct solder tip temperature with adequate (not excessive nor insufficient) heat transfer is essential for creating reliable solder joints.

A repeatable, stable rework process depends on time, temperature and operator technique. The contact time of the solder iron tip on the joint and the rework technique are dependent on operator skill and training. The actual temperature of the solder iron tip during the soldering process depends on the technology of the soldering iron. How well the solder iron recovers heat and puts the heat back at the tip, and the time the tip remains at the joint, determines the actual joint temperature.

Solder iron manufacturers are continually developing better performing irons. Each time a solder iron tip touches a joint, heat is conducted into the solder joint, requiring the heating element to power up to replace the heat loss. When the tip is removed from the joint, heat energy continues to flow for an instant as if the mass of the solder joint was still present. Subsequently, the tip temperature can overshoot.

For many sensitive components the overheating could damage components. On the other hand, a solder application on a heavy mass through-hole component on a multilayer board, requires all the power available from the heating element to transfer the necessary heat.

Work by Tsang [2] evaluated lead-free hand soldering rework with Sn4Ag and Sn3.8Ag0.7Cu cored wire on lead-free 20mil (0.5mm) pitch QFP (NiPd coated), 1206 chip (pure tin coated) and SOIC20 (NiPd coated) and PLCC44 (NiPd coated) on Immersion Silver and NiAu surface finish FR4 boards. The variables investigated were the lead-free rework alloy (Sn4Ag or Sn3.8Ag0.7Cu), the solder iron tip temperature (260°C (500°F), 316°C (600°F), 371°C (700°F), 427°C (800°F)), the flux volume (small or large flux solder rework core size) and the rework atmosphere (air or nitrogen).

The solder iron tip temperatures of 260°C (500°F) and 316°C (600°F) were found to be unacceptable for rework with either Sn4Ag or Sn3.8Ag0.7Cu cored solder wire with a minimum solder iron tip temperature of 370°C (700°F) solder iron tip temperature suggested without a measurable change in cycle time compared with tin-lead hand solder wire rework.

Increasing the volume of flux used tended to give better hand soldering results for lead-free rework for both lead-free alloys. Nitrogen was found to benefit the rework process in terms of cosmetic appearance of the lead-free reworked solder joint especially when reworking the 20mil (0.5mm) pitch QFP component.

The lead-free solder iron tip temperature suggested by Tsang [2] was similar to results found by Tsunematsu [4] on lead-free cored wire rework evaluations with SnAgCu. The lead-free solder spreading times for SnAgCu were found to be similar to SnPb when a solder iron tip temperature of 380°C (716°F) was used. The tip temperatures are slightly higher than those normally used for SnPb hand soldering rework. As the rework hand soldering temperatures are usually localized around the lead frame tip, the temperature of the component will not rise in the same way as for convection hot air gas rework such as for BGAs/CSPs, so only very temperature sensitive components will most likely be affected by this increase in tip temperature.

During NEMI rework evaluations [5] on TSOP and 2512 chip components on a 14 layer Cu, 7 × 17 inch (18mm × 43mm), 93mil (2.4mm) thick FR4 board the soldering iron tip temperature for the lead-free SnAgCu solder wire was 395°C (750°F) whereas the tin-lead Sn37Pb solder wire used a soldering iron tip temperature of 385°C (725°F). For the 135mil (3.5mm) thick FR4 board, the solder iron tip temperature for the SnAgCu solder wire was 395°C (750°F) whereas for Sn37Pb solder wire it was 385°C (725°F). Rework operators saw visual inspection differences, where the lead free reworked soldered joints appeared more cratered whilst the tin-lead reworked joint appeared more smooth and shiny.

Typical lead-free hand soldering temperatures, times and tips for SMT components are shown in Table 5.1. For boards with large thermal mass and/or greater than 93mil (2.3mm) thick, the lead-free soldering iron SMT rework temperatures and times shown in Table 5.1 may be higher.

In general the soldering iron tip temperatures used today for SnPb SMT hand solder rework are already high so there would not be a need to increase them significantly for lead-free SMT hand soldering rework. If soldering iron temperatures and times were increased for lead-free rework, temperature damage could occur resulting in Lifted pads, Blistering, Board Warping, Burning, Component Damage and Board/Component Delamination.

Based on experience from manufacturing, existing hand soldering equipment could be used for lead-free soldering but the equipment should be dedicated for lead-free rework to avoid issues with lead contamination. For lead-free SMT hand soldering, newer temperature controlled soldering

Table 5.1. Typical lead-free hand soldering SMT rework temperatures and times

Component type	Temperature [°C]	Contact Time [sec]	Recommended Tip for lead-free
< 1206 chip component	343°C to 371°C (650°F to 700°F)	3 to 5 seconds	Use tip the same size as the pad
> = 1206 chip component	371°C (700°F)	3 to 5 seconds	Use tip the same size as the pad
Gull wing and J-Lead type components	371°C (700° F)	3 to 6 seconds	Use tip the same size as the pad
Fine Pitch Components (less than 24mil pitch (0.6mm))	371°C (700° F)	3 to 6 seconds	Use tip the same size as the pad

irons would typically be recommended to avoid damage to components
and boards from excessive rework temperatures.

5.2.4 Soldering Iron Tip Life

When we consider the increased tip temperature and increased tin content
for higher tin lead-free rework solder alloys, the solder iron tip life can be
reduced which increases expense.

In general, most soldering iron tips are similar in composition (Figure 5.1).
The core usually is made from copper due to its high thermal conductivity.
Iron or other harder metals are used to help maintain the shape while preventing solder iron tip copper dissolution. The sides of the tip usually have
additional nickel plating, followed by chrome that prevents solder from
wicking away from the solder iron tip. Applying a coating of solder to the
tip, keeps the tip surface wetted and to serve as a thermal conduction path.

To increase solder iron life, soldering tip manufacturers can increase the
iron plating thickness, but iron has a relatively poor thermal conductivity,
which reduces thermal heat transfer efficiency so a balance needs to be
reached. The higher tin lead-free solder rework alloy also tends to erode
the iron plating on solder tips faster.

Re-tinning the solder iron tips is more critical for lead-free solder rework
and should be conducted frequently. Another variable affecting tip life is
the tendency of some operators to press harder while soldering with lead-free
alloys, thinking they will transfer more heat. While additional pressure does
not improve heat transfer, it does result in cracking and pitting of the iron
plating and increases degradation of the tip exposing the copper core and
shortening solder tip life. A high soldering iron tip idle temperature also

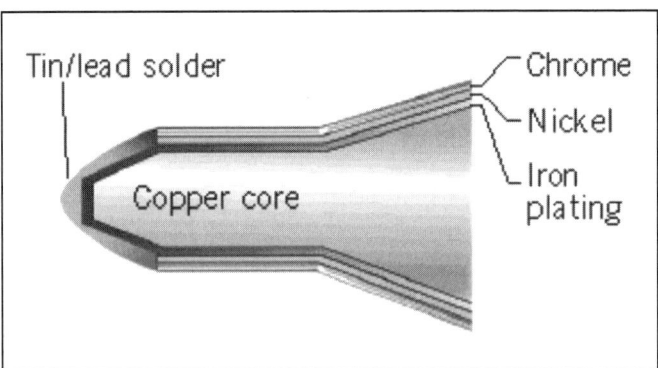

Fig. 5.1. Soldering iron tip construction (Courtesy O.K. Industries) [6]

reduces tip life in addition to the use of more active, aggressive rework fluxes.

5.2.5 Visual Inspection of Reworked SMT Solder Joints

The IPC-A-610D standard [7] can be used to determine whether the lead-free reworked solder joint is acceptable. It indicates that the primary difference between lead-free solder alloys and traditional tin-lead alloys is the visual appearance of the solder joint. The lead-free solder joint can appear duller or have a more grainy appearance than a tin-lead joint which is due to surface roughness of the high tin lead-free solder joint. A duller joint is not grounds for rejection of a lead-free or a tin-lead joint. Inspection of lead-free solder joints may show improper rejections of acceptable lead-free with texture and contact angles different from tin-lead.

5.2.6 Hand Solder Rework Reliability

There has been limited data on lead-free rework reliability but the data available indicates no real issues. Tsang [2] found lead-free SnAg or SnAgCu reworked NiPd 20mil (0.5mm) pitch QFP components had pull strength results equivalent or better than those for tin-lead reworked components on both immersion silver and NiAu board surface finish boards.

Reliability results from the NEMI Payette boards [5] for lead-free SnAgCu and tin-lead reworked TSOP components on 93mil (2.3mm) and 135mil (3.5mm) thick boards indicated no reliability failures after 6,000 accelerated thermal cycles from 0°C to 100°C.

5.2.7 Hand Solder Training

An essential part of lead-free hand solder rework is correct training as the increased temperatures and times can increase board and component damage. For good rework soldering we need to consider:

1. Rework Soldering Temperature
2. Rework Time
3. Rework Techniques (Procedures for Soldering and Reworking using Lead-free alloys)
4. Rework Tools (Consideration of the various shapes and configurations of the soldering iron tip for safe and cost effective rework)

Retraining would generally be recommended for hand solder operators and inspectors for lead-free.

5.2.8 Hand Solder SMT Rework Conclusions

Based on the results discussed, the choice of lead-free hand solder rework wire would be Sn3.5Ag, although there have been movements to Sn3Ag0.5Cu to be the same alloy as used for surface mount and wave soldering.

Typical soldering iron tip temperatures were found to be around 371°C (700°F) which was slightly higher than those used for tin-lead. The soldering tip would usually apply localized heating so that the component would be less likely to increase in temperature sufficiently to cause concern with this increased soldering iron temperature.

Good training would be needed to avoid excessive board and component damage. Soldering iron tip life could be degraded at a greater rate with lead-free high tin based solders with more care/attention needed by the operators for solder iron tip maintenance.

5.2.9 Future Work

Work would need to continue on development of the correct soldering iron tip temperatures and times for SMT hand soldering rework and development of hand soldering iron equipment with better temperature controls and longer soldering iron tip life. Lead-free hand soldering iron training also needs to be developed to ensure good soldering iron techniques and tools are used.

5.3 Lead-Free SMT Rework of BGA/CSP Soldered Joints

This section is broken down into the following areas:

- Alloy Choices
- Soldering temperatures
- Rework Nozzles and bottom/top heaters
- Component temperatures in relation to the J-STD-020 standard
- Adjacent component and board temperatures during rework
- Rework machine temperature tolerances
- BGA/CSP Rework Solder Joint Reliability
- Conclusions
- Future work

5.3.1 Alloy Choices

As the reflow alloy used for lead-free is SnAgCu, the rework material is SnAgCu solder paste with a melting temperature of 217°C or tacky/pasty flux. The decision on whether to use solder paste or tacky/pasty rework flux is usually dependent on the component type. For large I/O BGA parts, solder paste is typically used whereas for small CSP components pasty/tacky flux is used due to the difficulty in printing solder paste onto the fine pitch component or board locations.

5.3.2 Soldering Temperatures

Current BGA/CSP rework equipment can be used for lead-free rework operations but there are developments that need to be conducted for BGA/CSP rework equipment for lead-free processing as there is a need for equipment rated to higher rework temperatures which can keep the ΔT between the solder joint and top of the component as low as possible. The usual ΔT for BGA/CSP rework is around 15°C to 25°C between the solder joint and the top of the BGA/CSP component. This ΔT is far larger than a conventional hot air convection reflow oven (2°C to 3°C) as the component temperature stays hotter for longer in a rework machine.

Due to these reasons, the lead-free BGA/CSP rework solder joint minimum target temperature is usually 230°C peak to reduce the component top temperature as much as possible especially during BGA/CSP rework on large, thick lead-free soldered boards.

5.3.3 Rework Nozzles and Bottom/Top Heaters

With conventional hot air rework equipment, which is the most commonly used equipment in the industry, there are differences from the rework manufacturers in the type of rework nozzles and type of bottom rework heater plates and top nozzle heaters used. This affects the lead-free temperatures on the reworked component as well as the temperatures on adjacent components and the board.

In NEMI lead-free rework evaluations on the HP Yunque test board [8], it was found that the bottom-side heater was not uniformly distributing convection heat along the board during rework of a Mictor connector. The location of the Mictor component was near to the board edge of a 45cm (18 inch) × 30 cm (12 inch) board. The same results were found for two thicknesses of the test board (1.6mm (63 mil) and 3.5mm (135mil)). The Mictor component was at a location which was not directly over the bottom heater plate. This resulted in insufficient heat into the board from the bottom side convection heaters requiring excessive top nozzle heat which gave a large ΔT of 29°C between the solder joint and component top.

A follow on investigation reworking the Mictor connector on a "Resized" HP Yunque test board which had been cut in half, showed a reduced temperature delta observed between the solder joint and component top.

Work on the NEMI Payette board [9] showed that for lead-free rework, bottom side heater set points needed to be elevated compared to SnPb rework profiles to minimize the top nozzle heater set points which would over heat the top of the package beyond its IPC/JEDEC J–STD-020C standard package temperature limitations [10]. The lead-free bottom side heater set point were approximately 50°C higher than tin-lead bottom side heater set points.

The lead-free rework profiles were successfully developed for 2.3mm (93mils) and 3.5mm (135mils) thick NEMI Payette boards. The profiles were, on average, generated after multiple attempts through a two month time span. Some redesign of the bottom side rework machine heaters could be performed for better heat transfer. It was also found that adjacent component temperatures were exceeding the solder joint liquidus temperature at 150mils (3.8mm) away from the reworked component in many cases.

The top nozzle redesigns investigated for lead-free rework were not found to be mature in terms of development. During lead-free rework profile development the follow guidelines would typically be recommended:

(i)   Peak target temperature: 230°C (min. for solder joints) and 255°C (max. for components)

(ii) Time above solder joint liquidus: 45 to 90 seconds, and potentially over 90 seconds if not possible to keep to the 90 second limit

(iii) Linear ramp to peak temperature to help minimize ΔT across the component body

(iv) For thermocouple attachment to measure temperature, connect thermocouples to at least two solder joint and two package body locations (center and corner locations for both solder joint and component body)

It was difficult to minimize top package temperature while allowing a sufficient minimum peak solder joint temperature with adjacent and bottom side components undergoing a secondary reflow. More development would be needed for better bottom-side rework machine heat distribution to reduce top nozzle heat needed for rework machines especially on larger sized boards (30cm (12 inch) × 45cm (18 inch)).

5.3.4 Component Temperatures in Relation to the J-STD-020 standard

Various projects have evaluated the lead-free rework soldering temperatures for lead-free BGA/CSP components. In the NEMI 1999 lead-free project [11], lead-free PBGA256 components were reworked on a 63mil (1.6mm) thick soldered board. The reworked PBGA solder joint had a peak temperature of approximately 246°C and time over 217°C was 74 seconds. The top of the PBGA component reached 265°C and time over 217°C was 75 seconds. The temperature at 2.5mm (100 mils) away from the reworked component on the board was measured at approximately 193°C which would be within the limits acceptable to prevent secondary reflow of adjacent lead-free soldered components.

Other development rework projects included work on a network interface card comparing lead-free SnAgCu and tin-lead rework [12]. The rework equipment used was standard production BGA/CSP rework equipment. The board material was Epoxy FR4, 4 layer with Tg of 130°C to 140°C and thickness of 1.6mm (63mils).

The tin-lead PBGA rework profile had a solder peak temperature of 208°C, time over 183°C of 62 seconds and component body temperature peak of 219°C. The lead-free PBGA rework profile had a solder joint peak temperature of 240°C, time over 217°C for the solder joint of 48 seconds and component body temperature peak of 255°C with time over 217°C for the component of 54 seconds.

In work conducted by Gowda [13,14,15], lead-free Sn3.2Ag0.5Cu CSP46 0.75mm pitch packages were assembled with Sn3.5Ag0.7Cu solder paste on 1.6mm (63mil) thick FR4 NiAu surface finish boards and reworked with lead-free SnAgCu solder paste and lead-free CSP components. The peak temperature of the center solder joint during rework was 243°C with time over reflow of around 90 seconds. The top of the component reached a temperature of 262°C. An adjacent component at 4.5mm (180mils) distance from the component being reworked did not exceed 160°C.

Lead-free rework was also conducted on lead-free NiPdAu leadless micro-leadframe (MLF) 0.5mm pitch MLF48 components assembled with lead-free Sn3.5Ag0.5Cu solder paste on OSP coated 93mil (2.3mm) thick boards. This showed rework profiles with a 245°C peak at the solder joint and a component peak top surface temperature of 267°C. A location on the board 2.5mm (100mils) from the exterior of the rework nozzle reached 185°C peak temperature whilst an adjacent component 6.3mm (250mils) from the reworked component reached a maximum of 151°C temperature.

In work by Tsang (2), lead-free BGA256 (SnAgCu) and QFP208 (NiPd) components were soldered onto NiAu and Immersion Silver FR4 boards with SnAgCu solder paste and then hot air gas reworked at two different solder joint peak temperatures of 230°C and 250°C. The time over 217°C (60sec or 90sec), the use of different atmospheres (nitrogen or air), and the number of reworks (1x or 2x) were varied. The use of nitrogen during rework tended to give a more shinier solder joint and reduced the tendency for voiding. There was no difference between the 60 second or 90 second over reflow temperature in terms of solder joint quality. Results conducted on components undergoing 2x rework were not as good as those for 1x rework. The minimum rework joint peak temperature of 230°C gave similar results to that for 250°C peak.

In work by Sethuraman [16] on a surface mount test vehicle of 93mil (2.3mm) thick with a 8inch (20cm) × 8 inch (20cm) board dimension and high Tg FR4 board material (170°C) with 6 layers, the lead-free SnAgCu 0.5mm reworked CSP had a joint temperature of 231°C compared with a component top temperature of 265°C with a ΔT of 34°C. This compared with a first pass convection reflow oven profile for the same component showing a solder joint temperature of 247°C peak and a component top temperature of 250°C peak. The lead-free SnAgCu CSP84 component had a component dimension of 7mm × 7mm with 0.5mm ball pitch and component body thickness of 1.3mm. The component body volume was 66mm^3 which according to J-STD-020C [10] would have a temperature rating of 260°C peak.

Rework evaluations on the iNEMI Payette reliability test board [9] was done on 93mil (2.3mm) and 135mil (3.5mm) thick, 7inch (18cm) × 17inch (43cm), High Tg FR4 (170°C) board with 14 copper layers. Rework was conducted with standard hot air convection rework machines and standard nozzles on uBGA256, PBGA544 and CBGA933 components. The same rework machine model was used for all uBGA and PBGA components. An upgraded machine model was used for lead-free CBGA rework which had increased bottom heater capacity.

Increasing bottom side board heating was found to be effective in reducing ΔT between solder joint and component body for all three components as less top nozzle heat was needed but there were increased instances of secondary reflow of adjacent components and bottom side components. It was found that the margin of error to maintain a lead-free minimum solder joint temperature of 230°C to 235°C with maximum body temperature not exceeding 245°C to 250°C, which was indicated in a previous revision of the J-STD-020 standard, was very tight for some of the best rework development engineers in the industry. As a result J-STD-020C [10] was released with updates to allow for a 1x rework temperature rating of 260°C peak for any area array component irrespective of body size and thickness to take into account the issue of excessive lead-free rework temperatures on components.

It was generally observed that there was a tighter process window for lead-free rework due to component temperature limitations with lead-free rework profiles approximately 35% longer than SnPb rework profiles.

Chen [17] conducted lead-free CSP rework and found a 25°C ΔT between the solder joint and component top. Donaldson [18] conducted rework on lead-free BGA socket and FCBGA components. The BGA socket with package dimensions of 45 × 50mm and 1.27mm ball pitch had a rework solder joint temperature of 237°C with time above 217°C of 98 seconds with the plastic package handle of the component reaching 260°C with time above 217°C of 150 seconds. The FCBGA component with package dimensions of 35 × 35mm and 1.27mm ball pitch had a rework solder joint temperature of 236°C with time above 217°C of 77 seconds with the component body reaching 253°C with time above 217°C of 123 seconds.

The following gives a summary of lead-free CSP/BGA rework component temperature profiling results with the ΔT between solder joint and component body:

Chen[17]:	25°C ΔT (CSPs)
NEMI lead-free rework [11]:	18°C ΔT (PBGA)
NEMI lead-free rework [9]:	12°C ΔT (BGA)
Gowda [13,14,15]:	21°C ΔT (CSP/MLF)
Furnanz [12] :	15°C ΔT (PBGA)
Sethuraman [16]:	34°C ΔT (CSP)
Donaldson [18]:	20°C ΔT (BGA)
Average ΔT:	21°C

For lead-free rework soldering as the solder paste used would be Sn3Ag0.5Cu with a melting temperature of 217°C, the minimum solder joint peak temperature will be between 230°C to 240°C with an average ΔT of 21°C. This would mean that the top of a BGA/CSP using a lead-free rework profile would reach 251°C to 261°C.

A note in J-STD-020C standard [10] indicates that components which are not rated to 260°C peak for lead-free reflow due to their large body size or body thickness still need to be capable of withstanding a 1x 260°C temperature rating to simulate the temperatures encountered during lead-free area array rework.

5.3.5 Adjacent Component and Board Temperatures During Rework

As already indicated the adjacent component temperature and board can also be affected by BGA/CSP rework with potential secondary reflow of adjacent components causing reliability issues. The typical minimum keep out component spacing for BGA/CSP components is usually 3.8mm (150 mils) distance or greater from lead-frame components to avoid localized secondary reflow of adjacent components during rework operations. Designers are reducing spacing between components to reduce real estate usage on the board but this increases concerns of adjacent component issues.

During NEMI lead-free BGA/CSP rework studies on the NEMI Payette board [9], increasing the bottom side heater settings for the rework machine was found to have an adverse effect on bottom side and adjacent components in terms of exceeding lead-free liquidus solder joint temperatures (> 217°C). For both CBGA and uBGA rework on the NEMI Payette board, the board temperature (150mils (3.8mm) away from the component) exceeded the liquidus temperature for SnPb and SnAgCu soldering. There was an issue when reworking the uBGA256 near to the CBGA at location U27 on the NEMI Payette board. The adjacent CBGA component underwent a partial double reflow reducing solder joint integrity. Kapton tape was used to shield the CBGA component during uBGA rework to try and reduce this issue but there was still heat conduction through the board which still resulted in early CBGA solder joint fails during reliability testing.

An experiment was conducted to better understand the thermal characteristics of adjacent heating for the CBGA at U27 on the NEMI Payette. The adjacent CBGA away from the reworked uBGA had a liquidus solder joint temperature even though the spacing between the two components was 15mm (0.65 inch). There was found to be a thermal gradient across the CBGA package with some locations for the CBGA solder joint above and some locations for the CBGA solder joint below the lead-free liquidus temperature. It was interesting to note that if the adjacent component was fully reflowed which was the case for an adjacent uBGA, there did not appear to be any issues in terms of open solder joints and early ATC reliability fails for the adjacent uBGA.

Lead-free FCBGA780 rework on a 93mil thick high Tg FR4 board with 8 layers using an OSP surface finish was investigated by Yoon [19]. The component was 29mm × 29mm with a 1.3mm thick substrate thickness with a 1mm pitch and Sn3Ag0.5Cu solder balls.

A standard rework nozzle was used during the initial rework evaluations with a ΔT between solder joint and top of component body of 30°C (235°C for joint, 265°C for component body). An experimental rework nozzle design was evaluated which contained a baffle nozzle plate added inside the nozzle to help to deflect more heat around the sides of the component which would not then pass through the center of the component body so much reducing the component body temperature. This had the effect of keeping the component body temperature below the 260°C J-STD-020C limit but the new baffle nozzle had a 57 mm × 57 mm nozzle opening for a 27 mm × 27 mm component body size. The keep out spacing for adjacent components from the rework component with the new rework baffle nozzle was 14 mm but the typically keep out spacing in production usually went down to only 3.8 mm (150mils) to prevent adjacent component rework. Thus the new rework nozzle would be good to use for this test vehicle but not for real life production.

Further studies would be needed to develop rework processes and tools such as heat shields to reduce the adjacent component temperature issue but designers would need to be made aware of these issues and take appropriate steps to account for this in their designs.

5.3.6 Rework Machine Temperature Tolerances

There is limited or no data available for rework equipment temperature tolerances and there are many factors to control and consider when discussing rework machine temperature tolerances. Some of the factors include:

- Rework Machine supplier and machine model number
- The number, type and size of bottom heaters
- The board size and thickness
- The position, type and size of components on the board and whether the board is a single sided versus double sided board.
- The bottom heater and top nozzle heater settings and wattage/power of these heaters
- The variations in rework practices and top and bottom nozzle heater airflow rate set points and sequences used
- The rework nozzle type (open versus baffle) and nozzle dimensions and distance between the nozzle and the board during rework.
- The rework profile shape and length of profile and the variation in heating and cooling rates
- The reduced ability to cool the board down quickly after rework compared to a reflow oven
- The orientation of the board in the rework machine
- The ability and time of the rework process engineer to define and develop the optimal rework profile to use

Considering the iNEMI rework project [5], there were joint temperatures of 230°C to 235°C peak with component temperatures between 240°C and 256°C peak with ΔT between 10°C and 21°C. These were results developed for thermocoupled boards where multiple profiling runs (5 to 50) were conducted under controlled test conditions over a period of time (up to 2 months) by experienced process engineers. In production, there may not be thermocoupled boards and if these were available, the rework profiles may need to be developed in hours or days with the ΔT likely to be much worse giving issues with excess component body temperatures.

5.3.7 BGA/CSP Rework Solder Joint Reliability

In general the lead-free solder joint rework reliability test results for BGA/ CSP joints has been fairly good. The concerns would be related to controlling excessive intermetallic growth rates due to excess peak soldering temperatures and time above reflow and avoiding any component issues such as delamination in the package from excessive soldering heat.

Furnanz [12] had no issues during rework of tin-lead or lead-free BGAs using X-ray inspection. The reworked boards passed ICT and functional testing and underwent reliability testing with no defects encountered. The reliability testing on the tin-lead and lead-free assembled and reworked boards included ATC testing (–40°C to 85°C for 1,000 cycles), unpackaged shock and vibration tests, temperature and voltage operating and non-operating tests.

Gowda [13,14,15] conducted rework on lead-free Sn3.2Ag0.5Cu CSP46 0.75mm pitch packages assembled with Sn3.5Ag0.7Cu solder paste on 62mil thick FR4 NiAu surface finish boards. The reliability testing of as-assembled and reworked lead-free CSP components was found to have no issues after 2,500 air-to-air thermal cycles from 0°C to 100°C.

Even if the reworked component was reliable there may be concerns for the reliability of adjacent components and the board itself if excess soldering temperatures were used. As already indicated during uBGA rework in the iNEMI rework project [9], the reliability was found to have decreased for an adjacent CBGA near to the reworked uBGA component after ATC testing as the CBGA component has undergone a partial reflow during rework of the uBGA.

5.3.8 Conclusions

The temperatures being experienced during rework for the BGA/CSP components were found to be higher than during first pass reflow with the peak temperature of the components during rework 15°C to 25°C above that of the solder joint. More development would be needed in this area.

The margin of error to maintain a lead-free minimum solder joint temperature of 230°C to 235°C with a maximum body temperature of 245°C-250°C was very tight during BGA/CSP rework with the J-STD-020C standard helping to address this (260°C). Adjacent component temperatures were being exceeded 3.8mm (150mils) away from the reworked component on the board in many cases which gave cause for concern for adjacent component temperatures and joint reliability. Rework equipment and nozzles were still in the process of development for lead-free with a need for more lead-free process manufacturing margin for rework.

Increased bottom-side heat and thermal uniformity were critical to bring the board up to temperature for rework. More emphasis would be needed to optimize lead-free rework profiles, rework machine development and machine repeatability. These were under investigation in the new iNEMI lead-free rework optimization project which is discussed in the next section.

Education would need to be given to customers, component suppliers and the industry in general of the temperature concerns being faced during BGA/CSP rework.

5.3.9 Future Work

While convection air BGA/CSP rework machines are still the main stay of the industry, the need to reduce ΔT between the solder joint and component body requires more development and has also been renewed interest in alternative BGA/CSP rework equipment using Lazer and Infra-Red heating techniques.

For lead-free rework it is necessary to:

- Have better thermal controls of the equipment
- Use topside thermal heat shrouds for certain equipment to reduce the amount of bottom side heat needed
- Have better bottom-side heat distribution
- Consider newer rework nozzle designs
- Have tighter tolerances and more repeatable rework systems

While increasing bottom side board heat would reduce the amount of top nozzle heat needed, the main issues would be the affect on the board laminate material (in terms of reliability and warpage) and the bottomside components and adjacent component temperatures.

For nozzle redesigns, adjusting the heat airflow from the nozzle through the components can reduce the ΔT between the solder joint and the component body and try to reduce the risk of adjacent component reflow. For baffle style rework nozzles, although it reduced the ΔT between the solder joint and component top, the main issue would be larger keep out spacings around the component.

A follow on iNEMI lead-free rework optimization project [20] was initiated which would evaluate and recommend best practices, rework equipment requirements, impacts of adjacent component temperatures and best practice procedures for lead-free rework processing. This project which is separated into various sub-groups are discussed in the following sections.

5.3.9.1 INEMI Rework Adjacent Component Temperature Sub-Group

Due to the issues with the uBGA rework causing CBGA open solder joints on the NEMI Payette board [9], the use of   heat shrouds and heat shields (ceramic and other materials) are being investigated on the INEMI Payette board with an investigation to develop more uniform board heating. Other developments/tools under consideration include thermal modeling software to replicate the issue and simulate conduction versus convection heat into the reworked board and consideration and use of thermal imaging equipment to measure the heat distribution of rework machine bottom heaters which affects the heat into the reworked boards.

5.3.9.2 INEMI Rework Machine Temperature Repeatability and Tolerance Sub-Group

The two main parts to this project include getting an assessment of the current methods used by rework machine manufacturers and OEM/EMS companies to characterize the temperature repeatability and reproducibility of rework equipment to provide guidance on best practices to use and an assessment of current rework equipment in terms of rework machine temperature tolerances. This would involve the use of a modified rework machine temperature measurement equipment which was in commercial use. The project was divided into three stages:

Stage 1: The rework manufacturer to use the rework machine temperature measurement equipment with defined lead-free rework profile settings (fixed rework heat airflow rate, rework nozzle shape and distance from rework nozzle to board) to record temperature variations at specific thermocouple locations using a specific rework machine after multiple rework runs.

Stage 2: The same test as in Stage 1 to be conducted by the rework manufacturer on the same rework machine model but a different
rework machine.

Stage 3: The same test as conducted in Stages 1 and 2 but by an OEM/
EMS company with the same rework machine model but a different rework machine at the OEM/EMS company site.

A fixed number of repeatability runs would be done in each stage to determine average and standard deviation of time above liquidus, peak temperature and time with 5°C of the component peak temperature to determine
the gage repeatability for temperature measurement of the rework machine
using the same machine setpoint temperatures.

The gage repeatability data from each of the three stages would be assessed
to provide feedback to the rework machine manufacturers and improve
rework machine temperature repeatability and tolerances.

5.3.9.3 INEMI BGA Socket Lead-Free Rework Sub-Group

As there had been limited work on lead-free BGA socket component rework, the iNEMI Payette board would be redesigned to include a BGA
Socket component with dimensions of 45mm × 50mm. Rework of this
component would be followed by ATC reliability testing.

5.4 Lead-Free Pin-Through-Hole (PTH) Hand Solder Rework

Small through-hole components are typically reworked using hand soldering irons whereas multi-pin through-hole connectors and components are
typically taken to a solder mini-pot/fountain for rework. This section will
concentrate on lead-free PTH component rework using hand soldering iron
equipment and is broken down into the following areas:

- Solder Alloy Choices
- Soldering iron temperature and time
- Holefill, Pad lifting and Design Considerations for PTH hand soldering
- Hand Solder PTH Joint Reliability
- Conclusions

5.4.1 Solder Alloy Choices

With SnAgCu, Sn3.5Ag or SnCu being used for first pass wave soldering either of these alloys could be used for hand soldering PTH rework. As SnAg would be used for SMT hand soldering this would also be used for PTH hand soldering.

5.4.2 Soldering Iron Temperature and Time

Developmental work on lead-free PTH hand soldering has been limited. There would typically be similar soldering temperatures for SMT and PTH hand soldering but due to the additional affect of internal copper planes for PTH hand soldering dissipating more soldering heat into the board there may be a need to increase lead-free soldering iron temperatures and times.

Rework on a functional network interface card [21] on a few select component locations used a 371°C (700°F) soldering iron tip temperature. For another network card test vehicle board which had passive and daisy-chained components all lead-free hand-solder rework was also done with a 371°C (700°F) soldering iron tip temperature with some through-hole pads experiencing pad lifting and board damage. With a fully functional lead-free desktop four layer board, a variety of through-hole components were selected for rework. All hand solder rework was performed with a 371°C (700°F) soldering iron tip temperature with certain locations having solder shorts after rework and some broken and lifted pads.

Table 5.2. Typical lead-free PTH soldering iron temperatures and times

Component types	Soldering iron tip temperature [°C]	Contact Time [seconds]	Soldering Iron Tip Temperatures
Axial and Radial Components	371°C (700° F)	4 to 6 seconds	Soldering iron tip same size as board pad
DIPs, Connectors, and other Multi-Leaded Components	371°C to 399°C (700°F to 750°F)	5 to 7 seconds	Soldering iron tip same size as the board pad
Components soldered to the board with Multiple Thermal Copper Planes	Up to 427°C (Up to 800°F)	Contact time depends on the amount and type of preheat used	Soldering iron tip same size as the board pad. May need to apply heat from both sides of the board

Most of the rework problems were traced back to a lack of rework operator familiarity with lead-free solder rework. Use of a 427°C (800°F) soldering iron tip promoted a technique and rework speed similar to tin-lead rework. Some through-hole board pad lifting may be unavoidable due to the elevated rework soldering iron tip temperatures. Suggested temperatures/times/tips for lead-free reworked PTH (Pin-Through-Hole) components are shown in Table 5.2. Assemblies with board thicknesses of 2.3mm (93 mils) or higher and boards with multiple thermal planes may have higher temperatures and dwell times.

5.4.3 Holefill, Pad lifting and Design Considerations for PTH Hand Soldering

There are a variety of defects which can occur during lead-free PTH hand soldering. During PTH rework on a desktop board, there was some board pad lifting observed [21].

For a server test board of 2.3mm (93mil) thickness with 10 layers, a 427°C (800°F) soldering iron tip was used to rework the through-hole capacitor locations which gave an average hole-fill of only 26%. Pre-heating the board and component to 125°C followed by rework gave a 95% average hole-fill for the capacitors.

Components which had two layer direct connect and thermal relief pads had much better hole-fill than those with three and four layer connections. Four layer component locations demonstrated an average hole-fill of 8% more than the three layer locations because of the thermally relieved copper connections of the four layer locations.

5.4.4 Hand Solder PTH Joint Reliability

Solder joint reliability data has been limited. Work on a network interface card for PTH solder joint reliability [21] showed no failures with all boards passing functional tests. Reliability tests included using 1,000 cycles of temperature cycling from –40°C to 85°C and mechanical shock and vibration testing.

For the lead-free network test vehicle, the PTH hand rework boards went through 1,000 to 1,500 cycles of temperature cycling from –40°C to 85°C, and some boards went through mechanical shock and others through vibration with no issues.

Lead-free PTH reworked desktop boards also went through 1,000 cycles of temperature cycling from –40°C to 85°C, and other boards were tested through vibration and mechanical shock with no issues. Lead-free reworked server boards were also put through 1,200 cycles of temperature cycling and had no fails.

5.4.5 Conclusions

Although there had been limited work done on PTH lead-free hand solder rework, the testing done so far indicate soldering temperatures and time may be slightly higher than for lead-free SMT hand soldering rework. In certain cases, preheat of the board would be preferred. More work was needed to develop the rework processes with general soldering iron tip temperature guidelines from 371°C (700°F) to 427°C (800°F).

5.5 Lead-Free Pin-Through-Hole Mini-Pot Rework Soldering

This section is broken down into the follow areas:

- Solder Alloy Choices
- Mini-pot (Solder Fountain) Rework Machine Changes
- Soldering temperatures and times
- Component temperature
- PTH Holefill and Copper Dissolution
- PTH Mini-Pot Rework Solder Joint Reliability
- Conclusions
- Future work

5.5.1 Solder Alloy Choices

With SnAgCu, Sn3.5Ag or SnCu being used for first pass wave soldering either of these alloys could be used for lead-free Pin-Through-Hole Mini-pot machine rework. SnAgCu has been considered as the alloy to use as the main wave solder alloy has been SnAgCu but recent reports on the increased board copper pad and copper trace dissolution rate with the high tin containing SnAgCu solder have forced assemblers to consider other lead-free soldering alloys such as SnCuNi. This alloy along with similar types of alloy available from solder suppliers have been shown to help reduce the copper dissolution of the board pad [22,23] so would be considered as one of the alloys of choice at this time for lead-free mini-pot rework.

5.5.2 Mini-Pot (Solder Fountain) Rework Machine Changes

Mini-pot (solder fountain) rework machines can be retrofitted for lead-free soldering. There is a need to replace stainless steel parts in the machine and the rework nozzles with titanium parts which would not erode with the lead-free high tin containing SnAgCu or SnCuNi solders. In certain cases it may be easier to buy a new lead-free mini-pot machine than retrofit an existing piece of equipment which has been used for tin-lead rework. Preheating the board using a BGA rework machine or having preheaters as part of the mini-pot rework machine prior to the mini-pot soldering rework operation helps to ensure reasonable holefill for lead-free solder fountain rework. A rework machine with a preheater option is typically recommended.

5.5.3 Soldering Temperatures and Times

The NEMI Payette rework project [8] only covered DIP16 component rework on a 135mil (3.3mm) thick NEMI Payette board with 14 copper layers and a NiAu board finish.

For tin-lead solder, removing the wave soldered DIP16 components with no board preheat using the rework machine needed a pot temperature of 260°C with 6 seconds dwell time. For the lead-free SnAgCu soldered board, a pot temperature of 274°C was used with 15 to 20 seconds dwell time.

For replacing with a new DIP 16 component with no board preheat, for tin-lead solder a pot temperature of 260°C was used with 12 seconds dwell time. For the lead-free SnAgCu solder a pot temperature of 274°C was used with 30 seconds dwell time. Re-soldering (rework) was found to be much more difficult than first pass mini-pot wave assembly in terms of achieving good holefill.

Combining the times and temperatures using SnPb solder for first pass wave, removal and rework with no board preheat used, a pot temperature of 260°C was used with a total dwell time in the wave of 26 seconds to complete the three processes. Combining the times and temperatures using SnAgCu solder for first pass wave, removal and rework with no board preheat used, a pot temperature of 274°C was used with a total dwell time in the wave of 60 seconds to complete the three processes. It was evident that there would be a need for increased pot temperatures and dwell times for lead-free SnAgCu solder compared with tin-lead solder.

Cross-sectioning of the lead-free reworked DIP16 components showed copper barrel knee dissolution on the NiAu boards during the SnAgCu mini-pot rework as a result of the higher pot temperatures and dwell times and the higher tin content of SnAgCu compared with Sn37Pb solder.

Board preheat was evaluated for lead-free SnAgCu first pass and reworked PDIP16 components. The board pre-heat was performed using a BGA hot gas rework machine which preheated the topside of the board to 120°C followed by DIP rework over the solder fountain. The topside board temperature for a non-preheated board was only 80°C. Preheating of the board showed good hole-fill after the rework operation with reduced dwell times needed in the solder pot. For current SnPb production PTH rework processes board preheating prior to mini-pot rework would not typically be done.

As preheating of the board prior to lead-free soldering exhibited better soldering results, this would be one area of investigation in the iNEMI rework optimization study [20] as well as different component types including DIMM278 connector components on the INEMI Payette board and the use of different lead-free soldering alloys such as SnCuNi which could potentially reduce the rate of copper dissolution of the board pads.

5.5.4 Component Temperatures

Component temperatures during Lead-free PTH mini-pot rework have not been investigated thoroughly but during the NEMI rework project [5], the DIP16 component body temperatures were within specifications.

5.5.5 Holefill and Copper Dissolution

There are a variety of issues which can occur during lead-free Pin-Through-Hole rework soldering. As reported by Faure [24], with Sn3Ag0.5Cu alloy on a 2.3mm (93mil) thick lead-free wave solder test vehicle board, the copper from the PCB dissolved with increasing contact time with PTH copper barrel knee erosion after above 20 seconds of solder pot dwell time at a 265°C solder mini-pot temperature (Figure 5.2). This would compare with a pot temperature of 270°C to 275°C for SnCuNi alloy with a dwell time of 40 seconds before the same type of copper dissolution would occur for the same component and type of test vehicle.

The amount of copper dissolution varies as a function of the original barrel/pad copper thickness, the solder pot temperature and alloy (based on increased tin percentage typically increasing copper dissolution rate), dwell/contact time in the mini-pot and solder flow rate and the type of copper plating used on the PCB. Studies are ongoing with copper dissolution rate comparisons with various lead-free solder alloys and process optimization [22,23,24].

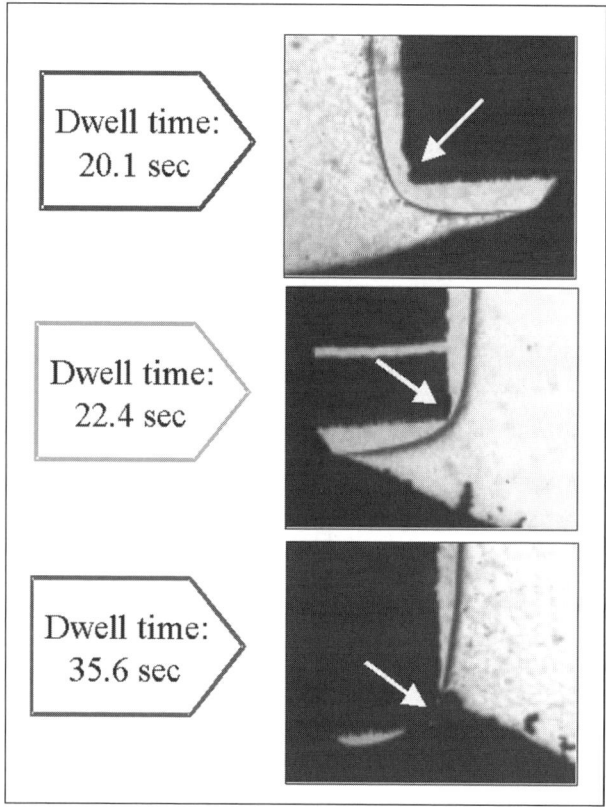

Fig. 5.2. Lead-free PTH solder fountain (mini-pot) rework showing cross-sections of copper knee dissolution as a function of mini-pot dwell time with Sn3.0Ag0.5Cu alloy [24]

With the NEMI Payette rework project [8], there were indications of insufficient holefill and increased copper dissolution at the barrel knee and bottomside copper traces during lead-free SnAgCu DIP16 rework.

5.5.6 PTH Mini-Pot Rework Joint Reliability

There is little or no data availability on reliability of lead-free PTH mini-pot reworked solder joints but it would be assumed that provided there was good holefill with no copper dissolution there should be no issues. The follow on INEMI rework optimization project [20] would investigate ATC reliability of lead-free PTH reworked joints on DIP16 and DIMM278 components.

5.5.7 PTH Mini-Pot Rework Conclusions

PTH component rework is still in development due to issues with increased solder mini-pot temperatures and dwell times using lead-free high tin containing solders which lead to issues such as copper dissolution and reduced holefill. Alternative alloys such as SnCuNi were being investigated to reduce the copper dissolution issue and more work also needed to be done in terms of the type of copper plating used by the board supplier to understand if this could also reduce the affect.

5.5.8 Future Work

The INEMI Rework Optimization project [20] would redesign the NEMI Payette board with one of the main purposes to develop and optimize PTH rework processes for DIMM278 as well as DIP16 components. Work would also investigate different lead-free rework alloys apart from SnAgCu to reduce the issue of copper dissolution such SnCuNi. The board thickness for the NEMI Payette board was 2.8mm (125mil) thick with OSP board finish with 14 copper layers. ATC reliability testing would be performed on the DIMM and DIP reworked boards from 0°C to 100°C up to 6,000 cycles.

Acknowledgments

The chapter author would like to acknowledge the work from the various persons, companies and consortia who contributed greatly to the data presented in this chapter.

References

1. Bath J, Sethuraman S (2002) Lead-free Solder Wire Rework Evaluation, Solectron Internal Report
2. Tsang M, Szymanowski RA (2002) Rework Processes for Lead-free Assembly, IPC APEX conference, San Diego, USA
3. Warwick M (1999) Implementing Lead-free Consortium Research, SMTA-I Conference, San Jose, USA
4. Tsunematsu Y, Tanigami M, Hirano M (1999) Evaluation of Pb-free Solders for Adaptability to Various Soldering Processes, IEEE
5. Bath J, Chu Q (2003) NEMI lead-free rework forum presentation, IPC APEX conference, USA
6. Curcio J (2005) Pb-free Process Control, Nepcon Shanghai, China

7. IPC-A-610D standard (2005) Acceptability of Soldered Electronics Assemblies

8. Bath J, Wageman M, Chu Q, Donaldson A, Ghalib N (2004) NEMI Lead-free and tin-lead rework development activities within the NEMI rework project, SMTAI conference, Chicago, USA

9. Gleason J, Reynolds C, Roubaud P, Kelly M, Lyjak K, Chu Q, Bath J (2005) INEMI lead-free rework evaluations, ECTC conference, Orlando, USA

10. J-STD-020C standard (2004) Moisture Sensitivity Level Classification for Non-Hermetically Sealed Packages

11. Walz M, Leahy T (2001) NEMI Lead-free BGA/CSP Rework Study Internal Report

12. Furnanz J, Tan SK, Bath J, Ruff C (2002) Manufacturing and Reliability Evaluation of a Lead-free Electronics Network Card, IPC Annual Meeting, New Orleans, USA

13. Gowda A, Srihari H, Primavera A (2001) Lead-free Rework Process for Chip Scale Packages, NEPCON East conference, Boston, USA

14. Gowda A (2002) Rework of Lead-free Surface Mount Components, IPC APEX conference, San Diego, USA

15. Gowda A (2002) Challenges in Lead-free Rework, Pan Pacific Conference, Hawaii, USA

16. Sethuraman S (2004) Solectron Lead-free SMT Reliability Test Vehicle Assembly and Rework Internal Report

17. Chen L, Yang A, Zhou W, Chen C (2000) Solectron Lead-free CSP Rework Internal Report

18. Donaldson A, Aspandiar R (2004) Hot Air Lead-free Rework of BGA Packages and Sockets, SMTAI conference, Chicago, USA

19. Yoon S, Bath J, Chu C (2004) Lead-free FCBGA assembly and rework evaluations, IPC APEX conference

20. Bath J (2006) INEMI Lead-free Forum Rework Presentation, IPC APEX conference, Anaheim, USA

21. Donaldson A (2006), Lead-free Hand Soldering PTH Rework, Circuit Assembly magazine

22. Hamilton C, Snugovsky P, Kelly M (2006) Lead-free mini-pot rework, SMTAI conference, Chicago, USA

23. Byle F, Jean D, Lee D (2006) Lead-free PTH rework optimization studies, SMTAI conference, Chicago, USA

24. Faure C (2005) Lead-free wave soldering studies, IPC/Soldertec Lead-free Conference, Barcelona, Spain

Chapter 6: Lead-Free Solder Joint Reliability

Jean-Paul Clech, EPSI Inc., Montclair, New Jersey, USA

6.1 Introduction

Lead-free solder joint reliability is a multi-faceted and challenging topic. Lead-free solders such as eutectic SnAg and SnBi have been used successfully in niche applications for many years. With the advent of no-lead (Pb) legislation, a multitude of soldering alloys has been proposed for mainstream electronic applications. The high number of lead-free alloy options remains a major factor slowing the development of reliability databases, test standards, acceleration factors and life prediction models. While the electronics industry has over fifty years of experience working with a single, main stream alloy - near-eutectic SnPb solder - the engineering community now faces the daunting task of qualifying product assemblies for an increasing variety of lead-free solders such as SnAgCu, SnAg, SnAgBi, SnBi, SnCu, SnZn alloys of near-eutectic or other compositions, some with known additive elements (e.g., nickel, cobalt, germanium) and others with proprietary compositions.

A significant amount of data, material properties, acceleration models and know-how has been developed for SAC387/SAC396 (SAC387 = Sn3.8%Ag0.7%Cu, SAC396 = Sn3.9%Ag0.6%Cu) solders - two alloys that were proposed by consortia in Europe and North America in the mid- to late 1990's. SAC305 (Sn3.0%Ag0.5%Cu) is now recommended by several industry associations. SAC205 (Sn2%Ag0.5%Cu) and SAC105 (Sn1wt%Ag0.5%Cu), and many other alloys, have also been implemented in real life applications. As of yet, long-term field experience is limited and material properties and/or reliability test data are not widely available for the latter alloys. Thus, full-fledged reliability programs are needed for a variety of alloys. For each and every alloy, the learning curve is quite steep and requires that close attention be paid to process, design and material details.

Establishing the reliability of lead-free assemblies for a given alloy first requires a thorough characterization of the physical and thermo-mechanical properties of both bulk solder and small solder joints. Such properties are critical to the development of physics-based acceleration factors and thus to the extrapolation of accelerated test results to real-life operating conditions. Acceleration factors are needed for soldered assemblies under mechanical loading (static creep, cyclical bending, vibration) and thermal conditions (thermal and power cycling). Secondly, extensive testing is required over a wide range of thermal and/or mechanical conditions - including vibration, shock and drop - so that the response of solders is understood under low to high stress or strain conditions.

This chapter begins with a summary of reliability test results for some of the mainstream lead-free alloys or board assemblies. A single chapter, let alone an entire book, cannot do justice to the subject of lead-free solder joint reliability. Rather than a review or a gap analysis on an alloy basis, this chapter presents a case study with the goal of presenting the type of data, material properties and analysis that are needed to estimate component attachment reliability for a given soldering alloy under thermal cycling conditions. One objective of this exercise is to illustrate the level of details that is required to develop stress/strain analysis models and estimate component attachment reliability in lead-free board assemblies. By definition, reliability is product and application specific and blanket statements about the reliability of a lead-free assembly - or the lack of it - need to be considered with caution.

6.2 General Trends

6.2.1 Lead-Free To Sn-Pb Comparisons

Modeling and thermal cycling test data for SAC assemblies with near-eutectic solder composition (SAC387/396) show that the ratio of SAC to SnPb lifetimes decreases with increases in one of the following three factors:

- Temperature swing (ΔT) or cyclic shear strains (Figure 6.1);
- Mean temperature (Figure 6.2);
- Dwell times at the temperature extremes (Figure 6.3).

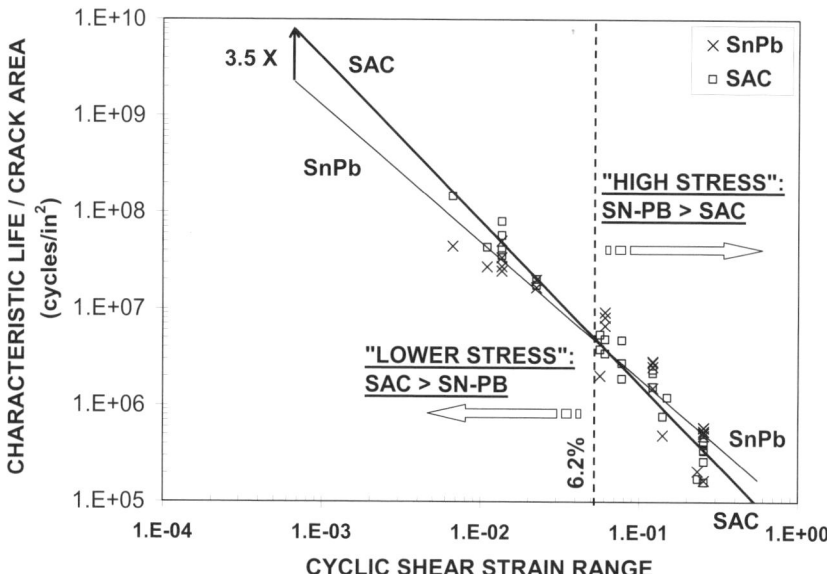

Fig. 6.1. Correlations of characteristic life scaled for solder joint crack area versus average cyclic shear strain in temperature cycling for standard SnPb and for SAC396 assemblies [1].

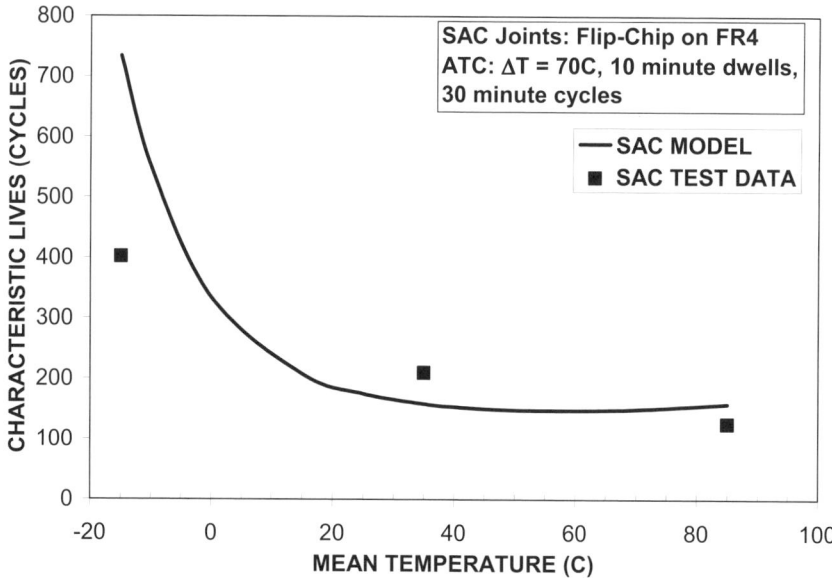

Fig. 6.2a. Mean temperature effect on flip-chip SAC387/396 solder joint life: test data from Schubert et al. [2]; predictive model (solid line) is from SAC387/396 hysteresis loop analysis [3].

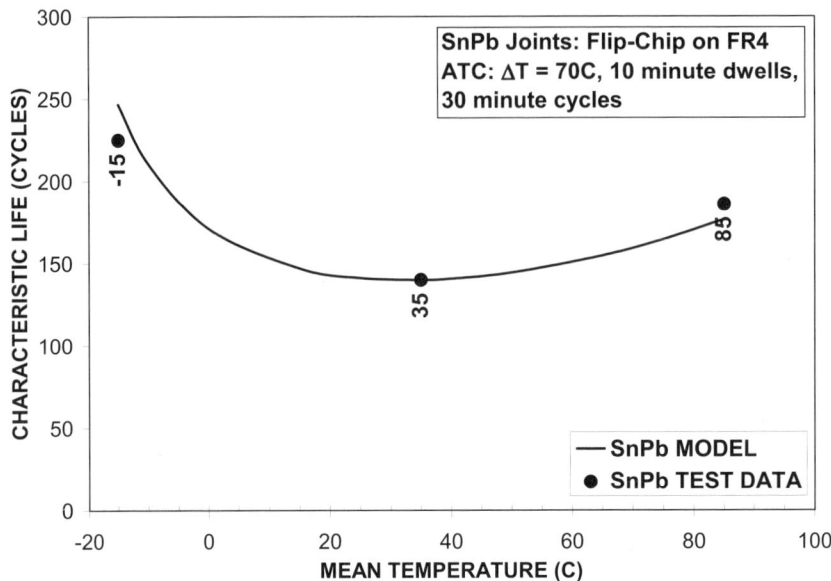

Fig. 6.2b. Mean temperature effect on flip-chip SnPb solder joint life: test data are from Schubert et al., 2003 [2]; predictive model (solid line) is from SnPb hysteresis loop analysis [3].

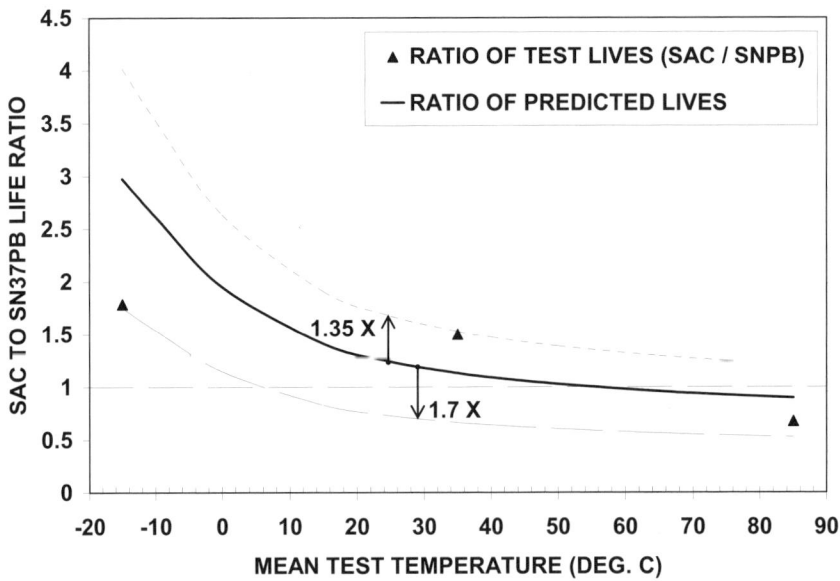

Fig. 6.2c. Ratio of SAC387/396 to SnPb characteristic lives from test data and life predictions in a) and b).

Fig. 6.2. Effect of mean cyclic temperature (average of cold and hot temperatures in a given thermal cycle) on: a) SnPb and b) SAC387/396 solder joint lives; c) ratio of SAC387/396 to SnPb lives.

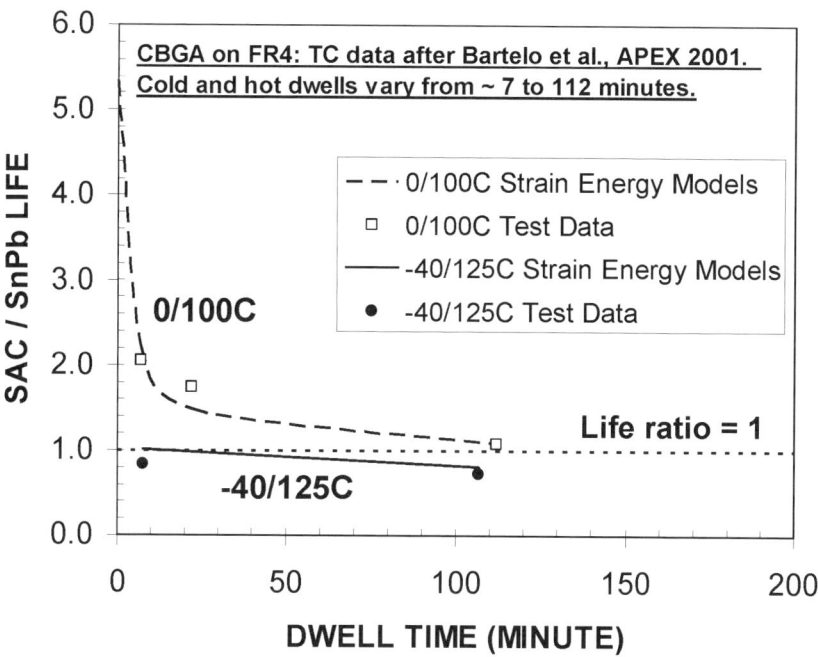

Fig. 6.3. Predicted and experimental trends of SAC to SnPb median life ratio as a function of dwell time in Thermal Cycling (TC) [3]. CBGA data points are from normalized failure distributions in Bartelo et al., 2001 [4].

The trends shown in Figures 6.1 through 6.3 are supported by both test data and modeling. A seemingly peculiar trend, shown in Figure 6.2b, is that SnPb life displays a minimum as the mean temperature increases. Additional test results showing a similar trend have been reported [3]. This is likely because, for a fixed temperature swing (ΔT), as the mean temperature goes up, SnPb solder softens faster than the rate at which the cyclic strain range develops. In other words, the stress and strain ranges evolve in opposite directions and the stress range - as given by the height of stress/strain hysteresis loops - decreases faster than the rate at which hysteresis loops widen.

Regarding the general trends, the ratio of SAC to SnPb life can be less than 1 under harsh enough test conditions. This does not imply that SAC assemblies are less reliable than SnPb assemblies since it has been observed that: 1) in general, SAC acceleration factors are larger than SnPb acceleration factors; and 2) correlations of SAC test data to applied shear strains under thermal cycling conditions have a larger slope than that of similar correlations of SnPb data (Figure 6.1). SAC solder joint life can be less

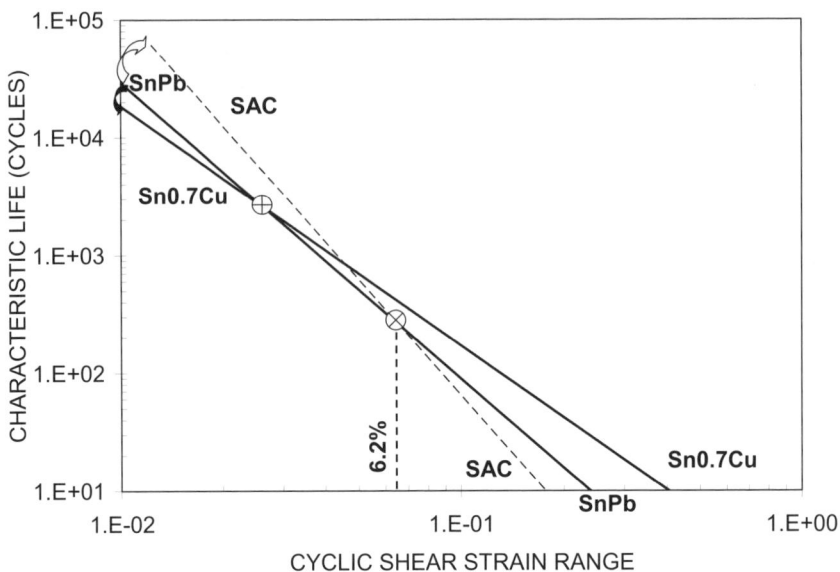

Fig. 6.4. Empirical correlations of characteristic lives and cyclic shear strains for bare chip assemblies. Trend-lines are based on accelerated thermal cycling data for Sn0.7Cu, Sn37Pb and near-eutectic SAC387/396 flip-chip assemblies [1].

that SnPb life under harsh enough conditions or high enough strains but may be longer under milder conditions or at lower strain levels. The above trends indicate that there is no simple lifetime ratio between SAC and SnPb solder joints. Their relative performances are highly product and application-dependent and one cannot infer from a single test or test data alone whether SAC life in the field will be better or worse than SnPb life. In the absence of long-term field returns, the only way to estimate field reliability is by extrapolating test results to field conditions using acceleration factors. The alternative, not extrapolating accelerated test data, is a procedure that is sometimes described as "taking a chance"!

The data in Figures 6.1 through 6.3 is for SAC387/396 solder joints. At this point in time, few similar syntheses of test results for other lead-free alloys are available in the public domain. One other trend-line was developed for cyclic lives of eutectic SnCu flip-chip joints versus average shear strains (Figure 6.4), with the SnCu line having a lesser slope than the SnPb line. This goes in the opposite direction of the SAC vs. SnPb trend-lines in Figures 6.1 and 6.4. Incidentally, although the SAC and SnPb trendlines in Figures 6.1 and 6.4 are based on entirely different datasets, they intersect at the same strain level of 6.2%. Regarding Figure 6.4, SnCu flip-chip joints display longer lives than SnPb joints under harsh enough test conditions but shorter lives under milder conditions. Again, one cannot rely on accelerated test data alone to assess and compare lead-free and SnPb solder joint reliability. In the end, what matters to end-users is not whether lead-free is more or less reliable than SnPb in test or in a generic sense but whether product reliability requirements are met.

6.2.2 SnAgCu (SAC) Alloy Comparisons

Lead-free reliability assessment remains a work in progress since new flavors of the SAC alloy family, and others, are being introduced on an on-going basis. SAC305, already popular in Asia, has been recommended by the IPC [5], with cost reduction and commonality being key driving factors. SAC105 and SAC205 and other new alloys are being used in products where the reduced stiffness and strength of SAC alloys with reduced silver contents provides for improved performance under mechanical shock or drop conditions. To this author's knowledge, no extensive reliability database is available yet for these new alloys (SAC105 and SAC205).

Preliminary data suggests that improvements under rapid mechanical loading may be associated with a life reduction under accelerated thermal cycling conditions. This is not surprising since failure modes and deformation mechanisms may be quite different. With a stiff solder alloy, higher forces are transmitted to intermetallic layers as well as to the pad interface with the board. Failures may then occur at the board to solder joint interface instead of in the bulk solder. A less creep resistant solder is more ductile and may absorb more of the impact energy under shock or drop conditions. Higher creep rates in a softer solder may also result in increased cumulative damage in the bulk solder during thermal cycling. However, a decrease in accelerated thermal cycling life does not necessarily imply insufficient product reliability. Here again, the development of acceleration factors, along with an estimate of their accuracy, is critical to the initial assessment of application-specific product reliability.

Comparisons of creep rates (Figure 6.5) and thermal cycling lives (Figure 6.6) for different SAC compositions indicate that, as expected, all SAC alloys are not equal. A plot of creep rates versus applied stress at different temperatures (Figure 6.5) for SAC105 and Sn3.5Ag0.75Cu specimens of identical geometry, and cast under similar conditions, shows that SAC105 creeps over one order of magnitude faster than Sn3.5Ag0.75Cu. The 45-degree plot of SAC405 versus SAC305 solder joint lifetime under thermal cycling conditions in Figure 6.6 shows all data points line up below the main diagonal. SAC305 solder joint life is an average of 35% below that of SAC405 solder joints.

Again, it is important to note that the plotted Land Grid Array (LGA) and flip-chip without underfill data in Figure 6.6 is for harsh conditions (–40/125°C and –55/125°C accelerated thermal cycling) and the ratio of SAC305 to SAC405 lives under other conditions is likely to be different. By how much can only be determined by conducting the appropriate tests or by extrapolating available test results to the use conditions of interest. The cost of such reliability studies is likely quite small compared to the price to pay for field failures, product returns and related litigations.

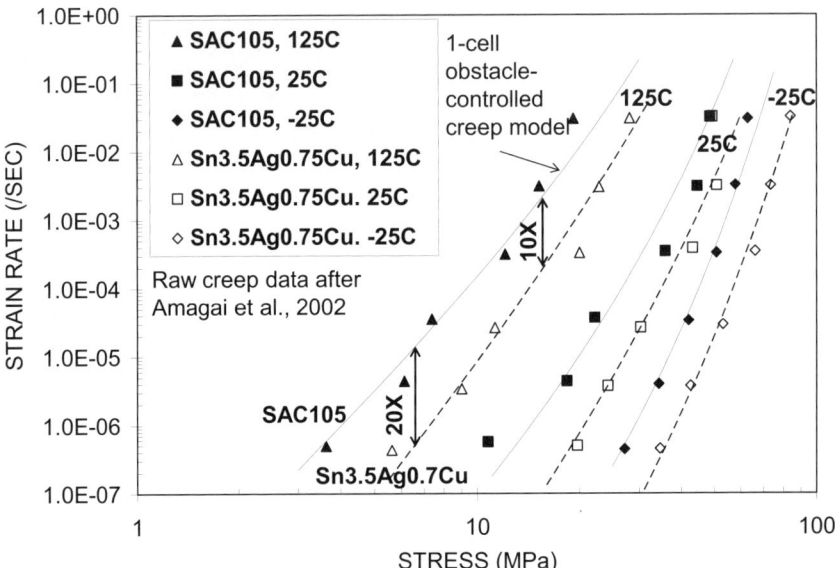

Fig. 6.5. Comparison of creep rates (strain rate vs. stress data at three temperatures) for SAC105 and Sn3.5Ag0.75Cu. Raw data is after Amagai et al., 2002 [6]; model lines are based on an obstacle-controlled creep model [7].

In summary, lead-free and SnPb solder joint lives are very much dependent on alloy composition, test and use conditions as well as board and package configurations. Reliability is not a material property and only refers to the ability of a product to survive expected loads for the intended design life of a particular product. As such, blanket statements to the effect that lead-free assemblies are more, or less, reliable than SnPb assemblies are of limited use when product characteristics, including alloy composition and use conditions, are not specified.

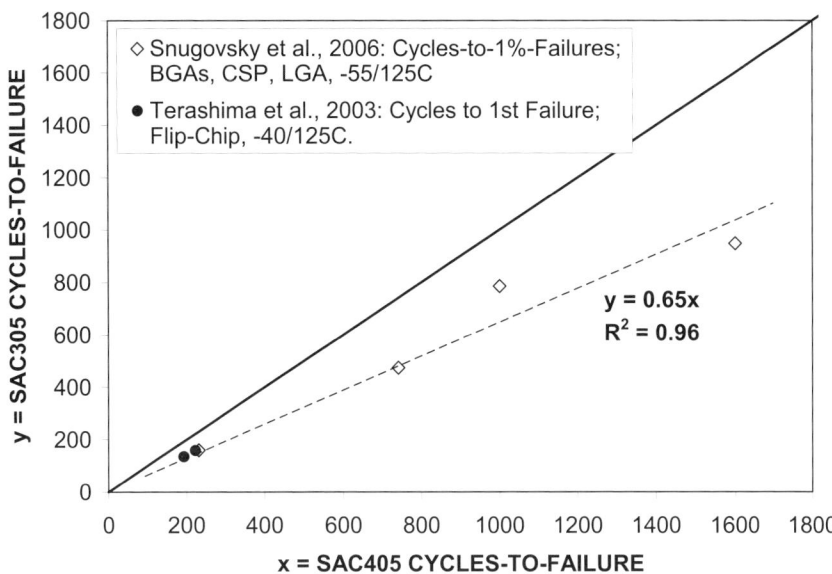

Fig. 6.6. 45-degree plot comparing SAC305 and SAC405 solder joint lives in flip-chip, BGA, CSP and LGA assemblies. SAC305 life (cycles to first failure or 1% fraction failed) is about 35% less than that of SAC405. Solder in reflowed joints was pure SAC305 or SAC405 (not mixed balls and paste). Raw data is from Terashima et al., 2003 [8] and Snugovsky et al., [9].

6.3 SAC Solder Joint Reliability Case Study

6.3.1 Problem Description

Consider a 676 I/O Plastic Ball Grid Array (PBGA) with SAC405 (Sn4%Ag0.5%Cu) balls, assembled on a FR-4 board using SAC396 solder paste. The package has a 17×17 sq. mm die that is 0.31-0.33 mm thick. The 676 I/O PBGA assemblies were tested under Accelerated Thermal Cycling (ATC) conditions (0/100°C) by a consortium of companies known as the Industry Working Group (IWG) [10]. The BGA ball and solder paste alloys were Sn4.0Ag0.5Cu and Sn3.9Ag0.6Cu, respectively. The peak reflow temperature was 235°C. The first goal of the present exercise is to illustrate extrapolation procedures by estimating the life of these PBGA assemblies under two sets of real life thermal cycling conditions:

- Condition # 1: 25° to 55°C, 1 cycle/day (1 cycle = 1440 minutes), 10°C/minute ramps (duration: 3 minutes), 717 minute dwell times at 25°C and 55°C.
- Condition # 2: 25° to 60°C, 4 cycles/day (1 cycle = 360 minutes), 10°C/minute ramps (duration: 3.5 minutes), 176.5 minute dwell times at 25°C and 60°C.

Use condition # 1 simulates diurnal (day and night) cycles in outdoors telecommunication cabinets. Condition # 2 relates to multiple daily cycles in portable computing equipment. Since the IWG tested the SAC PBGA assemblies under two ATC profiles - one with a short dwell time, the other with a longer dwell at 100°C - the second goal of this exercise is to show that predicted field lives are very similar whether test results are extrapolated from the short or the long dwell cycle. That is, using the results of two dissimilar tests - but with identical failure modes - predicted field lives are very much the same.

Field life projections will give a sense of how ATC cycles translate into years of field use for the 676 I/O SAC PBGA assemblies that were tested by the IWG. Similar field projections should be carried out for other lead-free solders, die sizes and components that are deemed critical in a given circuit board application with field-specific reliability requirements. In the absence of relevant field data for SAC assemblies, these reliability estimates cannot be validated with certainty. Rather, field reliability estimates such as those that are worked out in this case study serve as metrics for making go or no-go design decisions. The alternative - no field projections - leaves the door wide open for possibly significant reliability risks.

6.3.2 Input Data

Reliability analysis starts with gathering all the relevant input data: 1) a detailed description of the assembly geometry and material properties as required to build a finite-element model of the assembly; 2) thermal profiles for both accelerated testing and field use conditions; 3) failure times and statistics from ATC testing.

6.3.2.1 Geometry and Material Properties

All relevant package, board and joint dimensions and material properties are shown in Figures 6.7 through 6.9 and in Table 6.1. Figure 6.7 gives board thickness, die size, pad sizes and joint height. Figure 6.8 shows the maximum Distance to Neutral Point (DNP) at critical joints near the die periphery. These joints are expected to fail first because of the lower effective CTE of the package in the die area. Figure 6.9 also shows the thickness

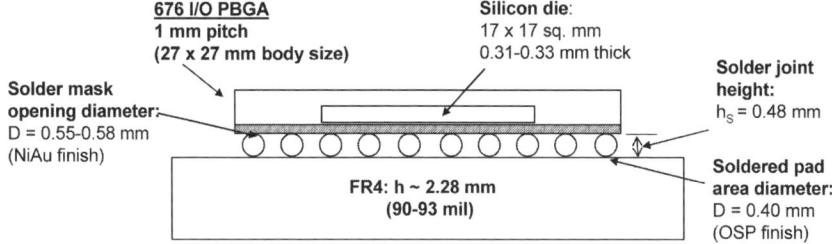

Fig. 6.7. Overall dimensions of 676 I/O PBGA assembly, including joint height and pad sizes. Information provided by Industry Working Group [10].

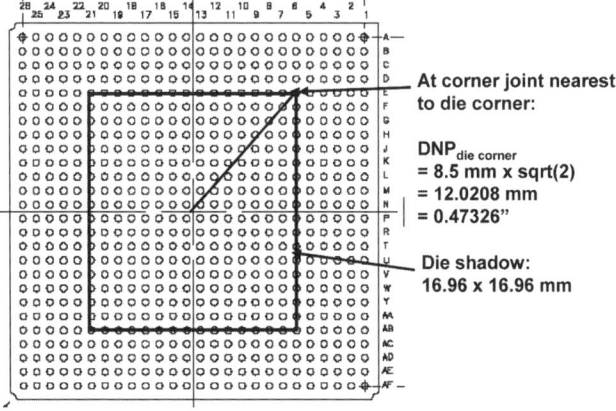

Fig. 6.8. PBGA676 footprint layout and die shadow showing Distance to Neutral Point (DNP) at critical joint nearest to die corner. Information provided by Industry Working Group [10].

of each layer in the schematic of the package cross-section in the die area. Elastic properties (Coefficient of Thermal Expansion or CTE, Young's modulus, E, and Poisson's ratio, ν) are given for each material in the multi-layer stack-up.

Material properties may also be required above and below the glass transition temperature of plastic or organic materials. As seen in Figure 6.9, no less than twenty parameters are required for the layer thickness and material properties of the stack of materials in the die area. The core properties of the FR-4 board laminate were given as: $E = 2208$ kg/mm^2, in-plane CTE (x & y) = 13 ppm/°C, $T_g = 180°C$. Accounting for the multi-layer stack of materials (laminate, copper planes, solder mask layers) in the test board structure (see Table 6.1), the nominal in-plane CTE of the test board was obtained as 14.8 ppm/°C. It is also important to note that, in many test boards as well as most product boards, in-plane CTEs depend on the board contents (copper planes, vias, filling materials, etc...) and are likely to differ by a few ppm/C in the x- and y- in-plane directions of the board.

All together, over three dozen parameters are required to describe the overall geometry and material properties of the 676 I/O PBGA assembly of interest. This huge set of parameters makes each component assembly unique. Except for silicon properties, which do not show much variability from one foundry to the next, caution should be used when searching for material properties of organic materials or constructions. For example, some five hundred different formulations are used for molding compounds across the electronics industry. Their material properties, rarely available in handbooks, are best measured on actual samples and can sometimes be obtained directly from the material suppliers.

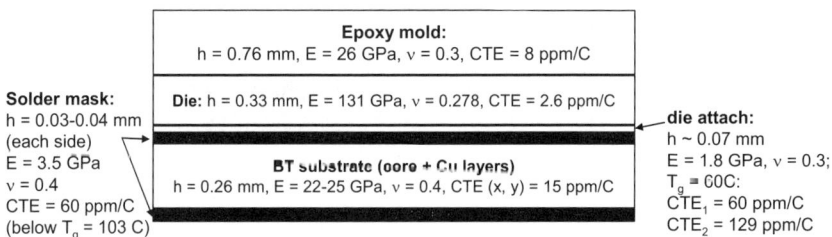

Fig. 6.9. PBGA material properties and thickness of each layer in die area. Information provided by Industry Working Group [10].

Table 6.1. Geometric parameters of PBGA assembly and test board structure (Source: Industry Working Group) [10].

Overall length dimensions	[mils]	[mm]
Molding Compound Length (at substrate insterface)	985.0	25.02
Die Length	667.6	16.96
Die Attach Length (at substrate insterface)	686.6	17.44
Substrate Length	1063.0	27.00

Overall thickness dimensions	[mils]	[mm]
Molding Compound thickness (above die)	30.1	0.76
Die thickness	12.9	0.33
Die Attach thickness	2.8	0.07
Substrate thickness	12.5	0.32
PCB thickness	93.0	2.36

Solder Joint dimensions	[mils]	[mm]
Solder Joint diameter (at 7 mils from PCB Cu pad)	26.8	0.68
Solder Joint height (Cu pad to Cu pad)	18.8	0.48
Solder Joint Pitch	39.4	1.00
Substrate pad diameter	23.0	0.58
Substrate side solder mask opening	18.0	0.46
PCB pad diameter	15.7	0.40
PCB side solder mask opening	21.5	0.55

Substrate thickness dimensions	[mils]	[mm]
Solder Mask over Cu (top)	1.2	0.03
Cu	1.0	0.03
Substrate Core	7.9	0.20
Cu	1.0	0.03
Solder Mask over Cu (bottom) - SMD	1.4	0.04

PCB thickness dimensions	[mils]	[mm]
Solder Mask over Laminate (top) - NSMD	1.8	0.05
NSMD Cu Pad thickness (top layer)	1.8	0.05
Laminate thickness	4.7	0.12
Cu	1.3	0.03
Laminate thickness	21.6	0.55
Cu	1.3	0.03
Laminate thickness	3.7	0.09
Cu	1.4	0.04
Laminate thickness	21.6	0.55
Cu	1.4	0.04
Laminate thickness	3.7	0.09
Cu	1.4	0.04
Laminate thickness	21.6	0.55
Cu	1.3	0.03
Laminate thickness	4.5	0.11
Cu (bottom Cu layer, no solder mask)	1.7	0.04

6.3.2.2 Thermal Conditions

Thermal cycling profiles and their parameters for accelerated testing and use conditions are shown in Figures 6.10 a and 6.10 b. For accelerated testing,

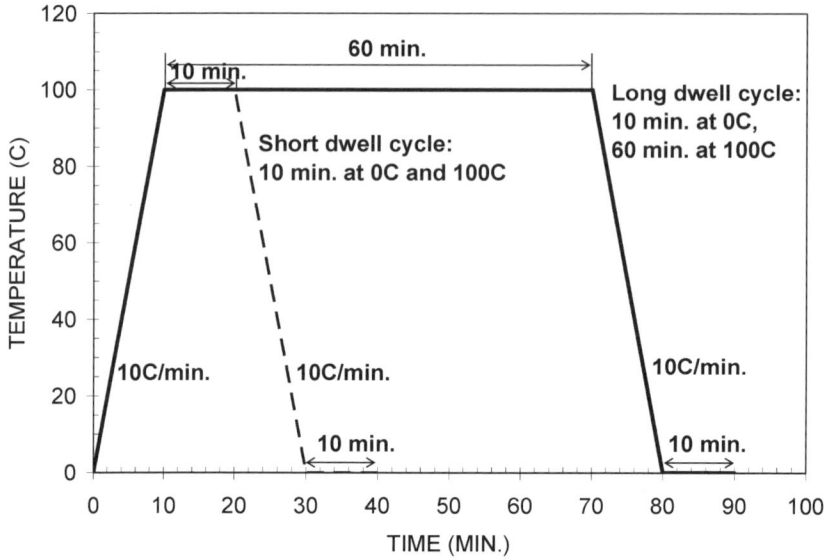

Fig. 6.10a. ATC profiles: short and long dwell cycles.

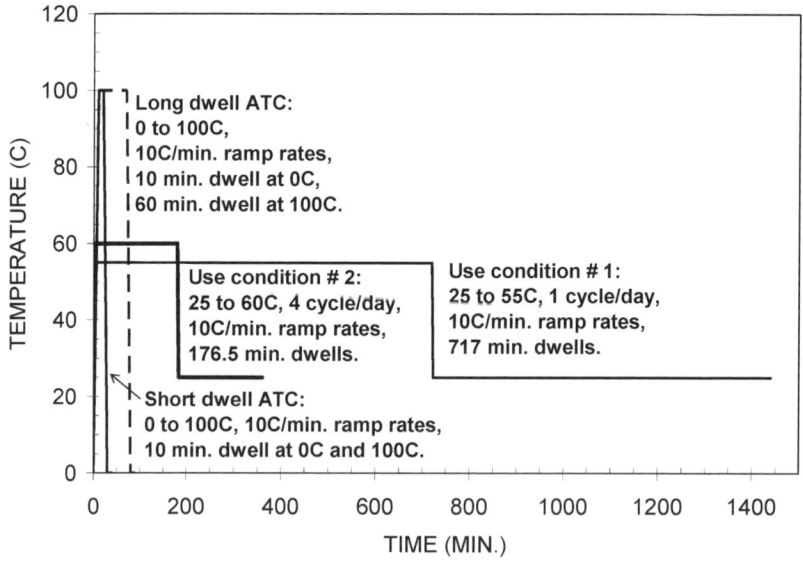

Fig. 6.10b. ATC and operating profiles for use conditions # 1 and 2.

Fig. 6.10. Thermal cycling profiles under a) test and b) use conditions.

the intent of the IWG project was to compare differences in failure modes, if any, when the dwell time at 100°C is changed from 10 minutes to 60 minutes. ATC ramp rates were 10°C/minute. One main objective of the present analysis, via the extrapolation of test results to use conditions # 1 and # 2, was to show that both accelerated test profiles lead to similar field life projections.

6.3.2.3 ATC Test Results

Figure 6.11 shows the distribution of failure cycles for the two test profiles on two-parameter (2P) Weibull paper format (fraction failed vs. number of cycles). Using three-parameter (3P) Weibull distributions, failure free cycles were obtained as: $N_{0,\text{Short Dwell}}$ = 3963 cycles and $N_{0,\text{Long Dwell}}$ = 3077 cycles for the "short" and "long" hot dwell cycles, respectively [3,10]. Failure mode analysis [10] showed that electrical opens occurred in bulk solder on the component side of the SAC solder joints near the die corners. The failure location was also confirmed by finite element modeling of the IWG PBGA assembly [11].

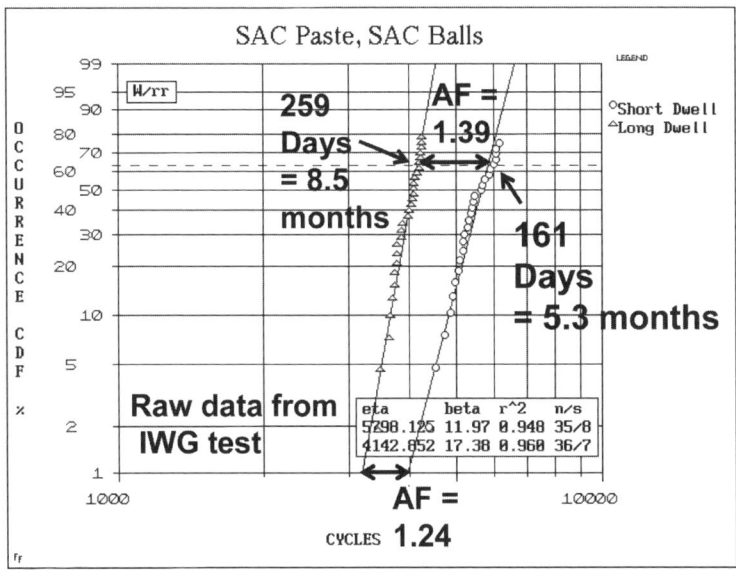

Fig. 6.11. 2P Weibull plot of failure cycles for SAC PBGA assemblies subject to ATC (0/100°C) with short (10 minutes) and long dwell (60 minutes) times at 100°C [3].

6.3.3 Reliability Analysis

The current section discusses the steps in reliability analysis: 1) the stress/strain analysis technique; 2) the choice of a constitutive model for SAC387/396; and 3) the empirical correlation of SAC387/396 thermal cycling data that is used to develop strain-energy based acceleration factors for SAC387/396 assemblies.

6.3.3.1 Stress/Strain Analysis

Acceleration factors can be obtained from the ratio of cyclic strain energies under test and use conditions using a one-dimensional Compact Strain Energy Model (CSEM) [3, 11]. The CSEM approach was selected for its ease-of-use and computational efficiency. The method was validated against various test sets as well as more sophisticated, and computationally more demanding, finite element based models [11]. Strain energies are given by the area of stabilized hysteresis loops that capture the rate and temperature-dependent stress/strain history of solder joints during thermal cycling. Stress/strain loops are computed for shear deformations driven by the global CTE mismatch between board and component, and for normal stresses and strains driven by local CTE mismatches between solder and the interconnected parts.

The stress/strain analysis follows the exact same procedures that were developed in the Solder Reliability Solutions (SRS) model [12-14] for SnPb assemblies except that temperature ramp rates are included in the lead-free analysis. Ramp rates are accounted for by performing a step-by-step integration of creep rate equations along the prescribed temperature profile. The methodology is identical to that developed by Hall [15-17] in the simulation of measured hysteresis loops for SnPb assemblies. The method applies to any shape of a periodic temperature profile, including trapezoidal, saw-tooth and sine profiles.

6.3.3.2 Constitutive Model for SAC387/396 Solder

The simplified constitutive model that was selected for SAC387/396 solder includes temperature-dependent elastic and minimum creep rate deformations. The elastic properties ($E = E(T)$ = Young's modulus; $G = G(T)$ = shear modulus; v = Poisson's ratio) used in the CSEM model are based on measurements of SAC specimens by Vianco et al. [18]:

$$E(T) = 2(1+v)\,G(T) \tag{1a}$$
$$G(T) = 20.24 - 2.635\ 10^{-2}\,T - 6.503\ 10^{-6}\,T^2 \quad \text{in GPa} \tag{1b}$$
$$v = 0.35 \tag{1c}$$

with the temperature T in degree Celsius. Poisson's ratio had very weak temperature dependence and is given as an average value over the temperature range –55/150°C.

The minimum creep rate equation that is used in the CSEM model is in the form of an obstacle-controlled creep model [7] that was fit to tensile creep data by Pang et al. [19] in the temperature range –40°C to 125°C (Figure 6.12). The model was shown (Figure 6.13) to apply to surface mount assemblies [7] by predicting the shear strength of SAC Chip Scale Package (CSP) type of joints at temperatures of 25°C, 75°C and 125°C [20]. Because of the variability of creep data with specimen size, geometry and configuration [21], it is critical to validate the choice of an adequate creep model by testing it against independent measurements on solder joints of real electronic assemblies.

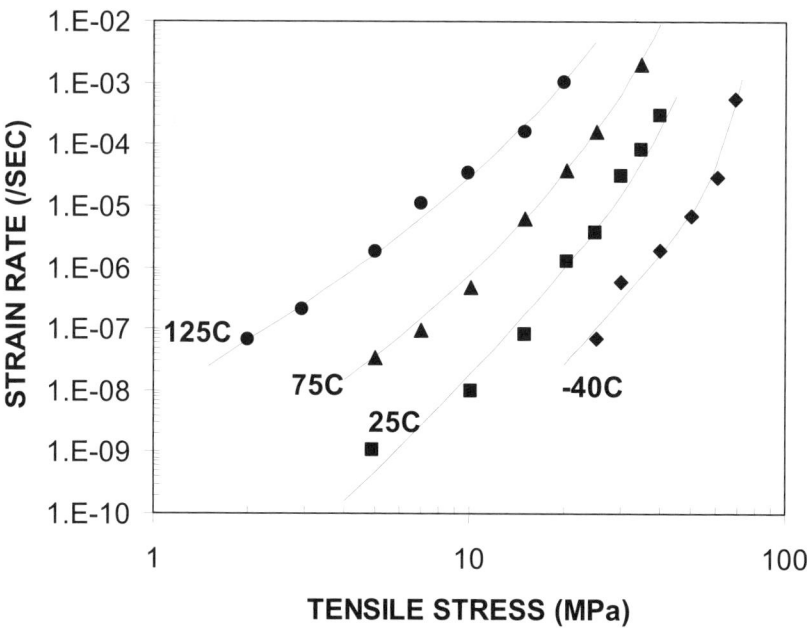

Fig. 6.12. Fit of two-cell, obstacle-controlled creep model to raw Sn3.8Ag0.7Cu tensile creep data [7]. The data is from Pang et al., [19]. The curve-fitted model lines (solid lines) are given as equation (2).

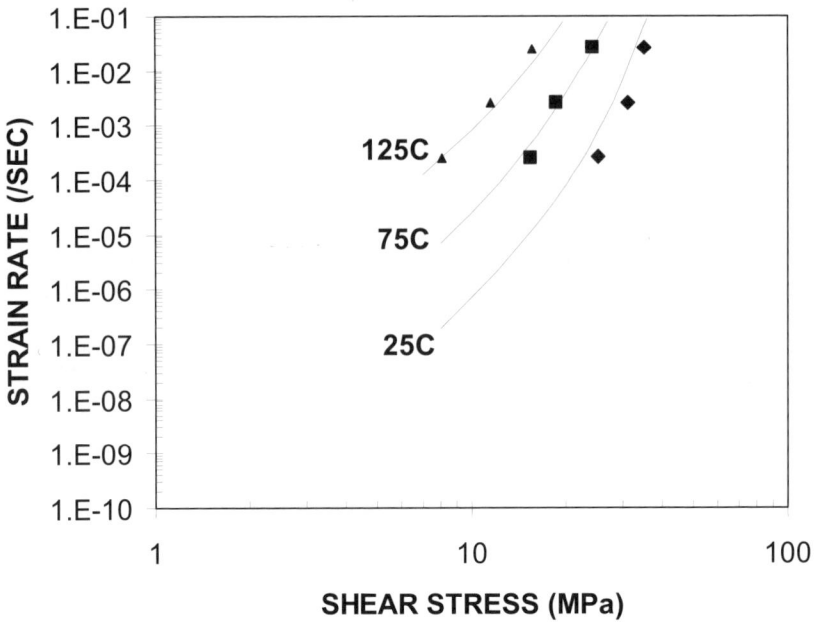

Fig. 6.13. Creep model validation [7]: fit of Sn3.8Ag0.7Cu model – equation (2) converted to shear in equation (3) – to CSP joint shear strength data from Pang et al. [20] at 25°C, 75°C and 125°C.

From curve fitting the original data [19] that was digitized and reproduced in Figure 6.12, the minimum creep rate equation in tension is given as:

$$\overset{\circ}{\varepsilon}(/\sec) = 5.0 \cdot 10^{-9} \sigma^{5.56} \cdot \exp\left[-\frac{3544}{T(K)}\left(1 - \frac{\sigma}{1280}\right)\right]$$
$$+ 6802\sigma^{3.02} \cdot \exp\left[-\frac{11050}{T(K)}\left(1 - \frac{\sigma}{181}\right)\right]$$

(2)

where the tensile stress σ has units of MPa and the temperature T is in degree Kelvin (K). Assuming that a Von-Mises yield criterion applies, creep rates in shear are:

$$\overset{\circ}{\gamma}(/\sec) = \sqrt{3} \times \left\{ \begin{array}{l} 5.0 \times 10^{-9} \left(\tau\sqrt{3}\right)^{5.56} \cdot \exp\left[-\dfrac{3544}{T(K)}\left(1-\dfrac{\tau\sqrt{3}}{1280}\right)\right] \\ \\ + 6802\left(\tau\sqrt{3}\right)^{3.02} \cdot \exp\left[-\dfrac{11050}{T(K)}\left(1-\dfrac{\tau\sqrt{3}}{181}\right)\right] \end{array} \right\} \tag{3}$$

where the shear stress τ has units of MPa.

A set of equations similar to equations (1) to (3) must be developed and validated for each and every new lead-free solder composition of interest. Examples of such constitutive models for some of the new alloys such as SAC105 and various SAC compositions with 0.05% nickel additives are given by Wiese et al. [21] and Dudek et al. [22].

6.3.3.3 Empirical Life Versus Strain Energy Correlation

For every solder alloy, acceleration factors are based on an empirical correlation of fatigue life versus some damage parameter. For SAC387/ SAC396 assemblies, thermal cycling failure data are correlated to cyclic strain energy given by the CSEM model (Figure 6.14). Similar correlations covering at least two orders of magnitude along the strain energy and life axis must be developed for each and every lead-free alloy composition.

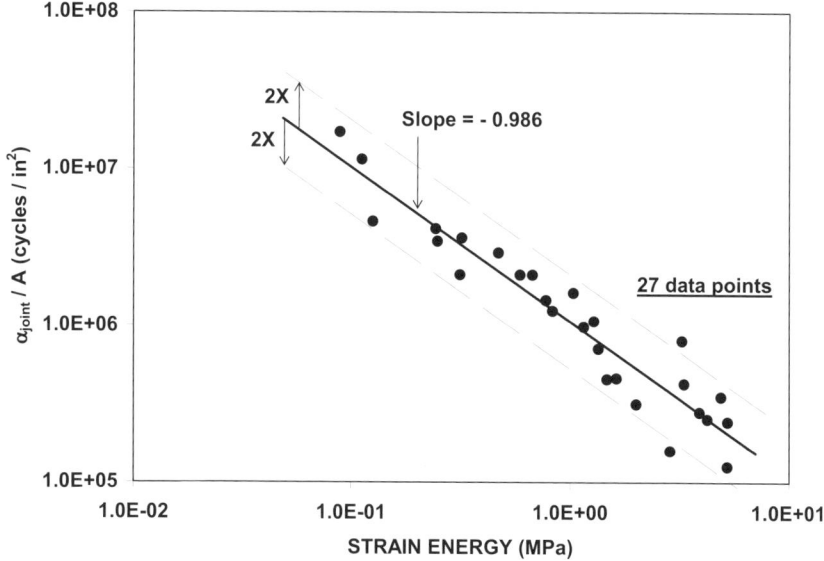

Fig. 6.14. SAC CSEM correlation: life scaled for crack area versus strain energy [3].

The equation of the best-fit-line through the data points in Figure 6.14 is:

$$\frac{\alpha_{JOINT}}{A} = \frac{C}{\Delta W^m} \qquad (4)$$

where α_{JOINT} is the characteristic life on a per joint basis (from Weibull failure distributions from a population of critical solder joints [12] as opposed to failure distributions for a population of components; critical solder joints are those joints that are likely to fail first, e.g., the eight corner joints of a peripheral chip carrier), A is the solder joint load bearing or crack area, C is a constant, ΔW is the cyclic strain energy density obtained from the area of stress/strain hysteresis loops, and the exponent m is close to 1 (m = 0.986).

The twenty-seven data points in Figure 6.14 cover about two orders of magnitude in each direction and the slope of the centerline correlation is close to -1. The spread of the data around the centerline is a factor of two times, which is typical of fatigue correlations. The slope of -1 is consistent with theoretical models and other strain-energy-based empirical correlations for thermal cycling fatigue of soft solders [3]. AFs(Acceleration Factors) are thus obtained as the ratio of cyclic strain energies (ΔW) under test and field conditions:

$$AF \equiv \frac{N_f(field)}{N_f(test)} = \frac{\Delta W(test)}{\Delta W(field)} \qquad (5)$$

where N_f s are cycles to failure and ΔWs are cyclic strain energies under test and field conditions.

The domain of validity of the CSEM model for SAC assemblies is dictated by the lower bound of strain energy values in the empirical correlation of Figure 6.14. Technically, the model and the ensuing AFs apply to use and test conditions with cyclic strain energies down to 0.1 MPa. In future work, and as more test data becomes available for SAC387/396 assemblies under milder stress conditions, we will check whether the correlation in Figure 6.14 holds at lower strain energy values. From previous comparisons of AFs from the CSEM approach, test results and finite-element models [11], the estimated accuracy of AFs obtained with the CSEM approach is about 30%. That is, calculated AFs or predicted field lives should be derated by a multiplicative factor of 0.7. Lastly it should be noted that a 50% accuracy on AFs is thought to be a reasonable and feasible goal [23].

6.3.4. Stress/Strain Cycles

Stress/strain hysteresis loops under ATC conditions with 10 or 60 minute dwells at 100°C are plotted in Figure 6.15. The ratio of loop areas is 1.26, that is, the predicted AF between the two test conditions is 1.26. The comparison of predicted and experimental AF in Table 6.2 indicates that the average experimental AF of 1.306 is 3.7% higher than the predicted AF of 1.26.

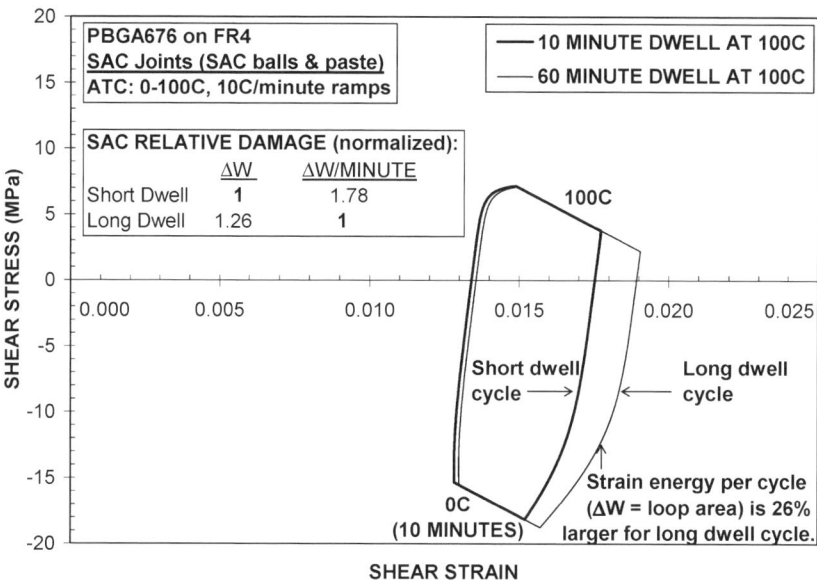

Fig. 6.15. ATC hysteresis loops under short and long dwell ATC conditions: shear stress vs. shear strain in SAC outermost corner joints under die shadow of 676 I/O SAC PBGA assembly on FR-4 [3].

Table 6.2. Comparison of predicted and experimental AFs for SAC PBGA676 assemblies under ATC conditions with 10 or 60 minute dwell times at 100°C [3].

	AF	**AF Calculation Method**
MODEL	1.26	Predicted as ratio of hysteresis loop areas in Figure 6.15.
TEST	1.24	Calculated as ratio of cycles to 1% failure in test (2P Weibull).
	1.39	Calculated as ratio of cycles to 63.2% failure in test (2P Weibull).
	1.288	Calculated as ratio of failure-free cycles in test (3P Weibull).
Test Average	1.306	Average of above three AFs in test.

The temperature and stress histories along the hysteresis loops of Figure 6.15 are plotted as a function of time in Figure 6.16. Stress relaxation proceeds at the fastest rate at the beginning of dwell periods (cold and hot). At 100°C, 50% stress reduction is achieved in the first ten minutes. Stress reduction slows down as time goes on and it takes another fifty minutes to reach 74% stress reduction at the end of the "long" dwell at 100°C. On the cold side, the percentage of stress reduction at the end of the 10 minute cold dwell is about the same (15 to 17%) for the two temperature profiles. In terms of damage rate, defined as cyclic strain energy per unit of time (or loop area over cycle duration in units of time), that of the "short" dwell cycle is 1.78 times larger than that of the "long" dwell cycle. That is, the "short" dwell cycle is more efficient at creating creep-fatigue damage in the SAC solder joints than the "long" dwell cycle.

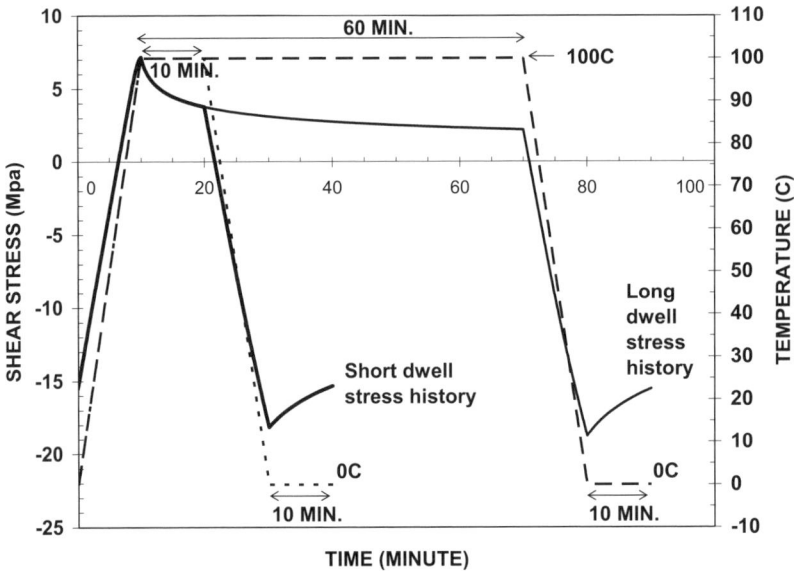

Fig. 6.16. Temperature and stress histories for SAC PBGA assemblies under "short" and "long" dwell ATC conditions [3].

Given that failure modes and mechanisms are similar in the two tests [10], and that the predicted AF is in excellent agreement with the experimental AF, the results of the two accelerated test profiles are consistent with each other. The whole experiment [10] and the hysteresis loop analysis lead to the conclusion that the "long" dwell time at 100°C is not necessary for the PBGA assembly of interest. However, long dwell thermal cycles are perfectly acceptable for organizations that can afford the time- and cost-premiums associated with extended test durations.

Hysteresis loops under use conditions # 1 and # 2 are shown in Figures 6.17 and 6.18, respectively, along with the loops for ATC conditions with short and long dwell times at 100°C. The hysteresis loops under ATC conditions are for PBGAs assembled on test boards with an approximate in-plane CTE of 14.8 ppm/C. The loops under use conditions # 1 and # 2 are for PBGAs assembled on "product" boards with CTEs of 14.8 (same as test board), 16 and 18 ppm/C. The hysteresis loops under use conditions increase in size with a larger CTE of the "product" board. This is as expected since a larger CTE mismatch on the product boards leads to shorter cyclic lives. Note also that, under the rather mild use conditions of interest, the loops only show limited stress reduction at the temperature extremes. However, and as expected, the loops for use conditions # 1 with a frequency of 1 cycle/day and dwell times of 717 minutes are wider than the loops for use conditions # 2 with a frequency of 4 cycles/day and dwell times of 176.5 minutes.

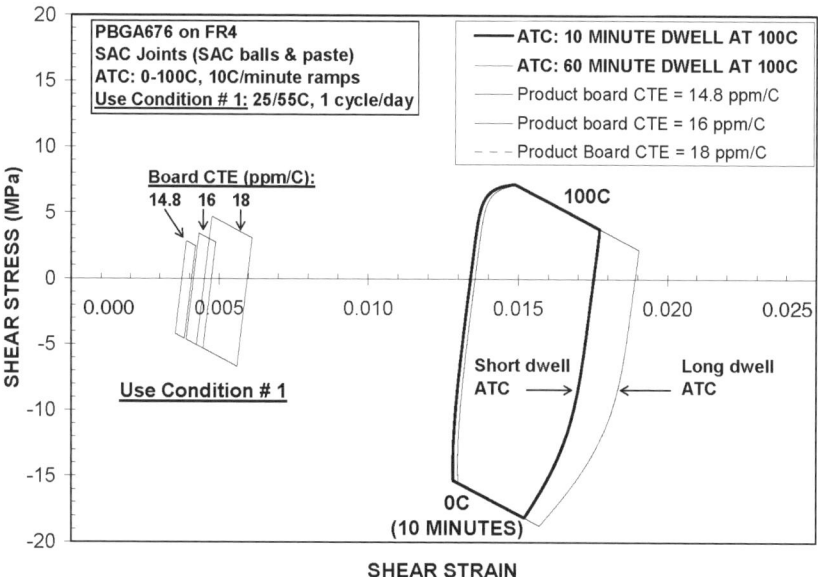

Fig. 6.17. Hysteresis loops for 676 I/O SAC PBGA assemblies under ATC (with test board CTE of about 14.8 ppm/°C) and Use Conditions # 1 (25/55°C cycle, 1 cycle/day) with FR-4 product board with different in-plane CTEs.

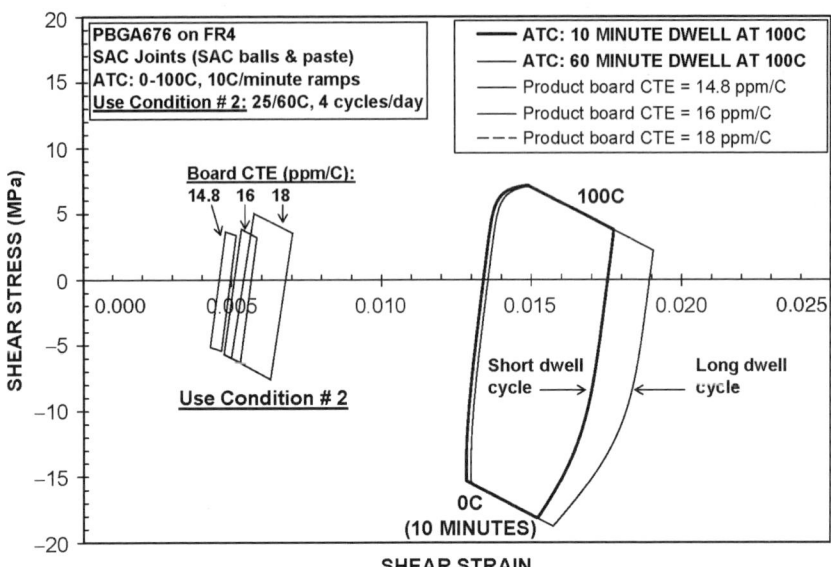

Fig. 6.18. Hysteresis loops for 676 I/O SAC PBGA assemblies under ATC (with test board CTE of about 14.8 ppm/C) and Use Conditions # 2 (25/60°C cycle, 1 cycle/day) with FR-4 product board with different in-plane CTEs.

6.3.5 Acceleration Factors and Field Projections

Acceleration factors and field life projections for use conditions # 1 and # 2 are given in Tables 6.3 and 6.4. AFs are calculated as the ratio of loop areas for use and test cycles in Figures 6.17 and 6.18. Field lives are given as cycles to failure (cycles to 0.1% failures when using 2P Weibull statistics or failure free cycles when using 3P Weibull statistics) and time-to-failure in the field in days or years. In the last column of both tables, years to failure are derated by 30% to account for an estimated error of ±30% on AFs. Field lifes for product board CTEs of 14.8, 16 or 18 ppm/C are obtained by extrapolation of test results for the two independent test conditions with dwell times of 10 or 60 minutes at 100°C.

Looking at the last columns in Tables 6.3 and 6.4, the projected field lives are very similar whether the extrapolations start with long or short dwell ATC data. This is quite a remarkable result, giving further confirmation that the two test profiles lead to very consistent failure data. The projected

Table 6.3. Acceleration factors and field projections for use conditions # 1 (25/55°C, 1 cycle/day): field life to 0.1% failures (2P Weibull analysis) and failure free times (3P Weibull analysis).

STATISTICAL ANALYSIS TYPE	PRODUCT BOARD CTE	TEST				AF	FIELD: USE CONDITION # 1: 25/55C, 1 CYCLE/DAY			
2P WEIBULL		ATC Dwell Time at 100C	2P Weibull parameters		Cycles to 0.1% failures	$\longrightarrow$	TIME TO 0.1% FAILURES			
			α (cycles)	β	$N_{0.1\%}$ (Test)		Cycles = $N_{0.1\%}$(Field) = AF * $N_{0.1\%}$ (Test)	Days = Cycles / 1 cycle/day	Years = Days / 365 days/year	Derated Field Life = 0.7* Years
	14.8 ppm/C (same as test board)	10 min	5798	11.97	3256	38.62	125753	125753	344.5	241.2
		60 min	4143	17.38	2784	48.75	135742	135742	371.9	260.3
	16 ppm/C	10 min	5798	11.97	3256	18.10	58921	58921	161.4	113.0
		60 min	4143	17.38	2784	22.84	63601	63601	174.2	122.0
	18 ppm/C	10 min	5798	11.97	3256	6.31	20534	20534	56.3	39.4
		60 min	4143	17.38	2784	7.96	22165	22165	60.7	42.5
3P WEIBULL (Failure-free time)		ATC Dwell Time at 100C			Failure Free Cycles: N_0(Test)	$\longrightarrow$	Failure Free Cycles: N_0(Field) = AF * N_0(Test)	Failure Free Days = Cycles / 1 cycle/day	Failure Free Years = Days / 365 days/year	Derated Failure Free Field Life = 0.7* Years
	14.8 ppm/C (same as test board)	10min			3963	38.62	153061	153061	419.3	293.5
		60min			3077	48.75	150018	150018	411.0	287.7
	16 ppm/C	10min			3963	18.10	71716	71716	196.5	137.5
		60min			3077	22.84	70290	70290	192.6	134.8
	18 ppm/C	10min			3963	6.31	24993	24993	68.5	47.9
		60min			3077	7.96	24496	24496	67.1	47.0

lifetimes based on two different test profiles are within 8% or 2% of each other when using 2P or 3P Weibull statistics, respectively. These conclusions hold regardless of the assumed in-plane CTE of the "product" board.

Tables 6.3 and 6.4 also show that AFs and field lives have a dependency on the product board CTE. Both decrease for product board CTEs that are larger than the test board CTE. In a previous study [3], it was shown that AFs had a weak sensitivity to the board-to-component global CTE mismatch as long as the test board and product board had the SAME in-plane CTE. However, and as expected, the results of the present study show that AFs vary when the test board and the product board CTEs are not the same. It is thus important that closed-form AF models account for the global CTE mismatch between boards and components.

Table 6.4. Acceleration factors and field life projections for use conditions # 2 (25/60°C, 4 cycles/day): field life to 0.1% failures (2P Weibull analysis) and failure free times (3P Weibull analysis).

STATISTICAL ANALYSIS TYPE	PRODUCT BOARD CTE	TEST			AF	FIELD: USE CONDITION # 2: 25/60C, 4 CYCLES/DAY			
2P WEIBULL		ATC Dwell Time at 100C	2P Weibull parameters		Cycles to 0.1% failures	$\rightarrow$	TIME TO 0.1% FAILURES		
			α (cycles)	β	$N_{0.1\%}$ (Test)		Cycles = $N_{0.1\%}$(Field) = AF * $N_{0.1\%}$ (Test)	Days = Cycles / 4 cycles/day	Years = Days / 365 days/year	Derated Field Life = 0.7* Years
	14.8 ppm/C (same as test board)	10 min	5798	11.97	3256	51.96	169195	42299	115.9	81.1
		60 min	4143	17.38	2784	65.60	182635	45659	125.1	87.6
	16 ppm/C	10 min	5798	11.97	3256	23.49	76476	19119	52.4	36.7
		60 min	4143	17.38	2784	29.65	82551	20638	56.5	39.6
	18 ppm/C	10 min	5798	11.97	3256	7.63	24836	6209	17.0	11.9
		60 min	4143	17.38	2784	9.63	26809	6702	18.4	12.9
3P WEIBULL (Failure-free time)		ATC Dwell Time at 100C			Failure Free Cycles: N_0(Test)	$\rightarrow$	Failure Free Cycles: N_0(Field) = AF * N_0(Test)	Failure Free Days = Cycles / 4 cycles/day	Failure Free Years = Days / 365 days/year	Derated Failure Free Field Life = 0.7* Years
	14.8 ppm/C (same as test board)	10min			3963	51.96	205936	51484	141.1	98.7
		60min			3077	65.60	201842	50461	138.2	96.8
	16 ppm/C	10min			3963	23.49	93083	23271	63.8	44.6
		60min			3077	29.65	91232	22808	62.5	43.7
	18 ppm/C	10min			3963	7.63	30230	7557	20.7	14.5
		60min			3077	9.63	29629	7407	20.3	14.2

Conclusions

As with SnPb assemblies, the assessment of lead-free solder joint reliability requires detailed, quantitative investigations of the thermo-mechanical response of solder alloys under a wide range of stress, strain and temperature conditions. Assuming that minimum workmanship requirements are met, lead-free solder joint reliability remains product- and application-specific. As illustrated in the case study written up in the second part of this chapter, assessing solder joint life for a given product board relies on a large number of parameters describing thermal conditions, board and component geometry and material properties. Significant progress has been made to quantify the thermo-mechanical behavior of SAC387/396 solders and board assemblies, including the development of test databases, life prediction models and acceleration factors. Similar efforts are required for each and every other solder alloy being considered for lead-free product assembly.

Acknowledgements

The author thanks the members of the Industry Working Group, in particular Jasbir Bath (Solectron), MJ Lee (Xilinx), Greg Henshall (Hewlett-Packard) and Keith Newman (Sun Microsystems) for their feedback on the manuscript and insightful discussions on the IWG test program.

References

1. Clech J-P (2004) Lead-Free and mixed assembly solder joint reliability trends, IPC Printed Circuits Expo/SMEMA Council APEX 2004 Conference, Anaheim, USA
2. Schubert A, Dudek R, Auerswald E, Gollhardt A, Michel B, Reichl H (2003) Fatigue life models for SnAgCu and SnPb solder joints evaluated by experiments and simulation, IEEE 53rd Electronic Components and Technology Conference, New-Orleans, USA
3. Clech J-P (2005) Acceleration factors and thermal cycling test efficiency for lead-free Sn-Ag-Cu assemblies, SMTA International (SMTAI) Conference, Chicago, USA
4. Bartelo J, Cain SR, Caletka D, Darbha K, Gosselin T, Henderson DW, King D, Knadle K, Sarkhel A, Thiel G, Woychik C (2001) Thermo-mechanical fatigue behavior of selected lead-free solders, IPC SMEMA Council APEX Conference, Anaheim, USA
5. IPC Technical Subcommittee of the IPC Solder Products Value Council Research Report (2005) Round robin testing and analysis of lead free solder pastes with alloys of tin, silver and copper
6. Amagai M, Watanabe M, Omiya M, Kishimoto K, Shibuya T (2002) Mechanical characterization of Sn-Ag-based lead-free solders, Microelectronics Reliability, Vol. 42, Issue 6, pp. 951-966
7. Clech J-P (2004) An obstacle-controlled creep model for SnPb and Sn-based lead-free solders, SMTA International (SMTAI) Conference, Chicago, USA
8. Terashima S, Kariya Y, Hosoi T, Tanaka M (2003) Effect of silver content on thermal fatigue life of Sn-xAg-0.5Cu flip-chip interconnects, Journal of Electronic Materials, Vol. 32, No. 12, pp. 1527-1533
9. Snugovsky P, Bagheri Z, McCormick H, Bagheri S, Hamilton C, Romansky M (2006) Failure mechanism of SAC 305 and SAC 405 in harsh environments and influence of board defects including black pad, SMTAI International Conference, Chicago, USA
10. Bath J, Sethuraman S, Zhou X, Willie D, Hyland K, Newman K, Hu L, Love D, Reynolds H, Koichi K, Chiang D, Chin V, Teng S, Ahmed M, Henshall G, Schroeder V, Lau J, Nguyen Q, Maheswari A, Cannis J, Clech J-P, Gibson C (2005) Reliability evaluation of lead-free SnAgCu PBGA676 components using tin-lead and lead-free SnAgCu solder paste, SMTA International Conference, Chicago, USA

11. Zhang R, Clech J-P (2006) Applicability of various Pb-free solder joint acceleration factor models, SMTAI International Conference, Chicago, USA
12. Clech J-P (1996) Solder Reliability Solutions: a PC-based design-for-reliability tool, Surface Mount International Conference, San Jose, USA
13. Clech J-P (1998) Flip-chip/CSP assembly reliability and solder volume effects, Surface Mount International Conference, San Jose, USA
14. Clech J-P (2000) Solder joint reliability of CSP versus BGA assemblies, System Integration in Micro Electronics SMT ESS and Hybrids Conference, Nuremberg, Germany
15. Hall PM (1984) Forces, moments, and displacements during thermal chamber cycling of leadless ceramic chip carriers soldered to printed boards, IEEE Transactions on Components, Hybrids and Manufacturing Technology, Vol. 7, No. 4, pp. 314-327
16. Hall PM (1987) Creep and stress relaxation in solder joints of surface mounted chip carriers, IEEE Transactions on Components, Hybrids and Manufacturing Technology, Vol. CHMT-12, No. 4, pp. 556-565
17. Hall PM (1991) Creep and stress relaxation in solder joints, Chapter 10, Solder Joint Reliability: Theory and Applications, ed. J. H. Lau, Van Nostrand Reinhold, pp. 306-332
18. Vianco PT, Rejent JA, Kilgo AC (2003) Time-independent mechanical and physical properties of the ternary 95.5Sn-3.9Ag-0.6Cu Solder, Journal of Electronic Materials, Vol. 32, No. 2, pp. 142-151
19. Pang JHL, Xiong BS, Low TH (2004) Creep and fatigue characterization of lead free 95.5Sn-3.8Ag-0.7Cu, 54th Electronic Components and Technology Conference, Las Vegas, USA
20. Pang JHL, Xiong BS, Neo CC, Zhang XR, Low TH (2003) Bulk solder and solder joint properties for lead-free 95.5Sn-3.8Ag-0.7Cu solder alloy, IEEE 53rd Electronic Components and Technology Conference, New-Orleans, USA
21. Wiese S, Roellig M, Mueller M, Rzepka S, Nocke K, Luhmann C, Kraemer F, Meier K, Wolter K-J (2006) The influence of size and composition on the creep of SnAgCu solder joints, 1st ESTC Conference, Dresden, Germany
22. Dudek R, Rzepka S, Dobritz S, Döring R, Keyssig K, Wiese S, Michel B (2006) Fatigue life prediction and analysis of WLPs with Sn98.5Ag1Cu0.5 solder balls, 1st ESTC Conference, Dresden, Germany
23. Pan N, Henshall GA, Billaut F, Dai S, Strum MJ, Benedetto E, Rayner J (2005) An Acceleration Model for Sn-Ag-Cu Solder Joint Reliability Under Various Thermal Cycle Conditions, SMTA International Conference, Chicago, USA

Chapter 7: Backward and Forward Compatibility

Jianbiao Pan, California Polytechnic State University, San Luis Obispo
Jasbir Bath, Solectron Corporation, Milpitas, California
Xiang Zhou, Solectron Corporation, Milpitas, California
Dennis Willie, Solectron Corporation, Milpitas, California

7.1 Introduction

In response to the European Union (EU) Restriction of Hazardous Substances (RoHS) and other countries' impending lead-free directives, the electronics industry is moving toward lead-free soldering. Total lead-free soldering requires not only lead-free solder paste but also lead-free printed circuit board (PCB) finish and lead-free component/packages. Transitioning tin-lead (SnPb) soldering to totally lead-free soldering is a complex issue and involves movement of the whole electronics industry supply chain. In reality, there is a transition period.

In the early transition phase, consumer electronics manufacturers wanted to convert their products to be lead-free quickly to comply with environmental regulations and avoid a marketing disadvantage. But some lead-free components/packages were not available because components manufacturers were slow in responding to the lead-free transition or there was insufficient demand initially. Thus tin-lead components were assembled with lead-free solder paste. This would be termed a forward compatibility situation.

In the late transition phase, many component manufacturers had migrated to lead-free production. Since the demand for tin-lead components was low, component manufacturers did not want to carry both SnPb and lead-free production lines due to the cost concerns. Therefore, some components such as memory modules are no longer being made available in SnPb finish. On the other hand, some products, such as servers, are exempt from the EU RoHS directive until or beyond 2010. Additionally products such as medical equipment, and military and aerospace products are not required to be lead-free. These products want to continue to be built with conventional SnPb solder paste because the reliability of SnAgCu – or lead-free – solder joints for these high reliability applications is still unknown. This scenario, soldering of lead-free components with SnPb paste, is known as the backward compatibility situation.

Table 7.1 summarizes the transition to total lead-free soldering. Tin-silver-copper (SnAgCu or SAC) solders have been considered to be the best alternative to SnPb solders for most applications. The most common alloys are Sn3.0Ag0.5Cu (SAC305) recommended by Japan Electronics and Information Technology Industries Association (JEITA) and IPC Solder Value Council and Sn3.8-3.9Ag0.6-0.7Cu recommended by iNEMI (InterNational Electronics Manufacturing Initiative) and the European consortium – BRITE-EURAM. Here Sn3.0Ag0.5Cu means 3.0% in weight Ag, 0.5% in weight Cu, with the leading element Sn making up the balance to 100% by weight.

Lead-free PCB finishes are used in both the backward compatibility and the forward compatibility assemblies. The most common lead-free PCB finishes include Organic Solderability Preservatives (OSP), Immersion Silver, Electroless Nickel Immersion Gold (ENIG), and Immersion Tin.

Table 7.1. Transition to Total Lead-free

	SnPb solder paste	Lead-free solder paste
SnPb components	Traditional SnPb soldering	Forward compatibility
Lead-free components	Backward compatibility	Total lead-free soldering

There are challenges in both forward compatibility and backward compatibility, especially for the BGA (Ball Grid Array)/CSP (Chip Scale Package) component for which this chapter will discuss. The microstructure and reliability data using SnAgCu BGA/CSP spheres with SnPb paste will be reviewed. The estimation of mixed solder composition liquidus temperature will be presented. The chapter then presents leadframe and chip components backward compatibility. Forward compatibility will be also briefly discussed. Finally, status of lead-free press-fit connectors will be presented.

7.1.1 Challenges to Backward Compatibility

Backward compatibility means lead-free packages/components attached to a printed circuit board (PCB) using SnPb solder paste. Since different package types have different metallizations, backward compatibility issues differ by package types. For BGA/CSP packages, the typical package metallization is a SnAgCu ball. For leadframe components such as Quad Flat Packages (QFPs) and Small Outline Integrated Circuits (SOICs), the typical lead-free component metallization is pure tin (Sn), Sn3.5Ag, Sn1.0Cu,

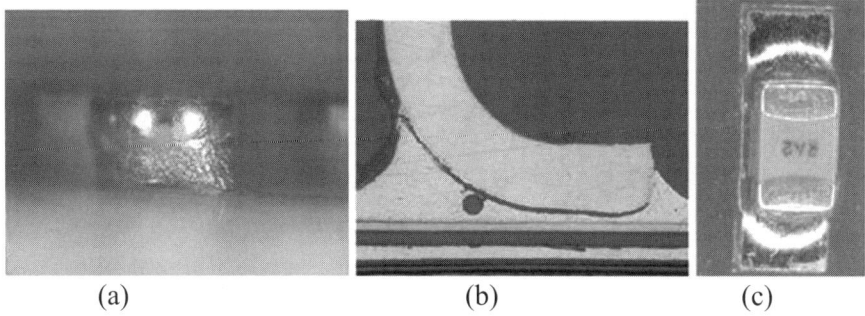

(a)   (b)   (c)

Fig. 7.1. Backward Compatibility for Three Package Types (a) SnAgCu BGA sphere with SnPb paste; (b) Leadfree QFP or SOIC with SnPb paste; (c) Pure Sn chip component with SnPb paste

Sn2-4% wt. Bi, NiPdAu or NiAu. For termination or chip components, the most common metallization is pure Sn. Fig. 7.1 shows backward compatibility for these three package types.

A schematic of BGA/CSP backward compatibility is shown in Fig. 7.2. The liquidus temperature of SnAgCu (SAC305 or SAC387) is between 217 to 221°C and the typical reflow peak temperature of SnAgCu solder paste is between 230 to 250°C. The liquidus temperature of eutectic SnPb is 183°C and the typical reflow peak temperature of eutectic SnPb solder paste is between 200 to 220°C. The question is what reflow profile should be used for backward compatibility assembly, a SnPb profile, a SnAgCu profile, or another profile?

If a SnAgCu profile is used, the SnAgCu solder ball will melt and the solder ball will self-align as shown in Fig. 7.3(a). But there are two issues. Firstly, the reflow temperature may be too high for other SnPb components on the same board or the board itself during assembly. Table 7.2 summarizes the component rating per IPC/JEDEC J-STD-020C. Secondly, the flux in SnPb solder paste may not function properly at such a high reflow

Table 7.2. Component Rating per IPC/JEDEC J-STD-020C

	Eutectic SnPb solder	SnAgCu solder
Liquidus temperature	183°C	217-221°C
Typical reflow peak temp	200 ~ 220°C	230 ~ 250°C
Component rating Per IPC/JEDEC J-STD-020C	225 +0/–5°C for large thick components (240°C for small & thin components)	245°C for large thick components (260°C for small & thin components)

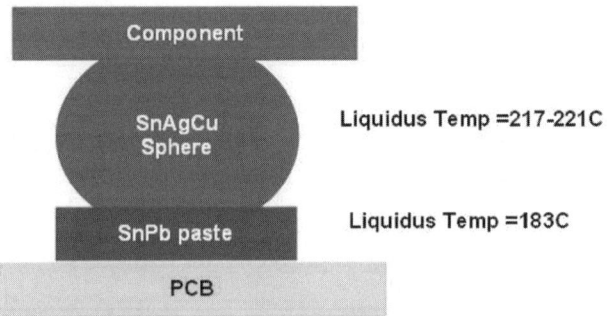

Fig. 7.2. A Schematic of BGA/CSP Backward Compatibility

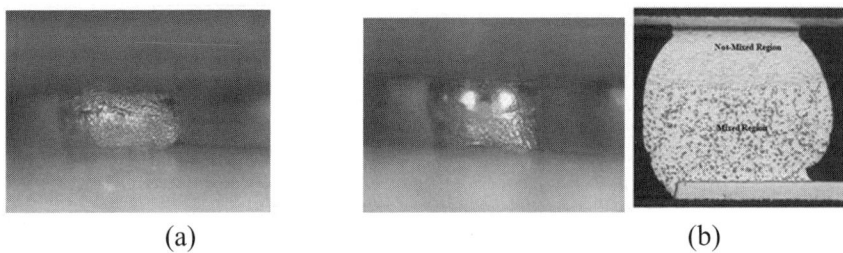

(a)   (b)

Fig. 7.3. Comparison of different reflow profiles in backward compatibility (a) using a SnAgCu reflow profile; (b) using a SnPb reflow profile

temperature. On the other hand, if the SnPb reflow profile is used, the SnAgCu solder ball will only partially melt and won't be self-aligned as shown in Fig. 7.3 (b). The incomplete mixing of solder and no self-alignment raise reliability concerns. Therefore, the key in BGA/CSP backward compatibility assemblies is to find the minimum reflow peak temperature to be able to achieve complete mixing of SnPb paste with lead-free components with good self-alignment.

7.1.2 Challenges to Forward Compatibility

Forward compatibility means tin-lead packages/components attached to a PCB using lead-free solder paste. In BGA/CSP forward compatibility shown in Fig. 7.4, a SnAgCu reflow profile is typically used. However, more voids were found in the forward compatibility solder joints [1, 2]. The greater voiding in the solder joints has become a reliability concern [3]. Another issue is that the high SnAgCu reflow peak temperature may exceed the maximum temperatures that the SnPb components are allowed to reach.

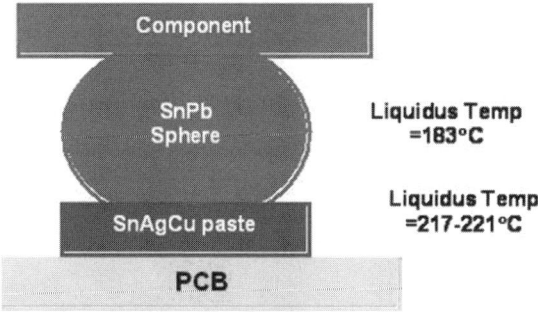

Fig. 7.4. A Schematic of Forward Compatibility

7.2 Reliability of BGA/CSP Backward Compatibility

7.2.1 Microstructure of Backward Compatible Joints

It is important to evaluate the joint microstructure of lead-free BGAs soldered with SnPb solder paste since the microstructure is a good indication of the solder joint reliability. The degree of mixing in backward compatibility assembly is expected to be a function of the reflow peak temperature and time above liquidus.

Grossmann et al. investigated various reflow profiles (peak temperature and time above liquidus) on the microstructure of the solder joint [4]. The package they tested was a PBGA200 with 13 mm × 13 mm component body size, 1 mm thick, 0.8 mm pitch, and 7 mm × 7 mm die size. The solder ball in the PBGA was Sn3.8Ag0.7Cu with a diameter of 0.5 mm and a height of 0.3 mm. The PCBs were made of glass epoxy FR-4 with a thickness of 1.58 mm (62 mil). The PCB finish was electroless Ni immersion Au (ENIG). The stencil used for solder paste printing was laser cut with aperture openings of 0.7 mm (28 mil) (1:1 to the pad size) and a thickness of 0.15 mm (6 mil). The solder paste was Sn36Pb2Ag, Type 3 with rosin-based no-clean flux. Table 7.3 summarizes the results of the work. Their results show that the SnPbAg solder paste interacted with SnAgCu ball even at a peak temperature of 210°C, but the solder ball was only partially mixed with the solder paste. The SnAgCu ball was fully dissolved when the peak temperature reached 217°C. The dendrites got smaller as the peak temperature increases. It should be noted that the melting point of Sn36Pb2Ag is 179°C compared with 183°C for the more common Sn37Pb solder.

Table 7.3. Summary of Grossmann et al. [4]'s Results

Peak temperature (°C)	Time above SnAgCu liquidus (sec.)	Time above SAC387 liquidus (sec.)	Results
210	62	–	SAC ball partially reacted with SnPbAg solder
217	66	–	SAC ball is fully molten, but the dispension of Pb is inhomogeneous, IMC is formed
218	82	6	
223	71	28	Completely mixed; the dendrites are smaller than that soldered at 217C
227	90	46	
233	76	43	Completely mixed; homogeneous distribution of the Pb-rich phase; fine IMC.
246	100	73	

Bath et al. used a FBGA676 I/O, with a pitch of 1.0 mm on their backward compatibility study [5]. The package body was 27 mm × 27 mm in body size containing a die of 17 mm × 17 mm in size. The solder ball diameter was 0.6 mm. The solder ball composition was Sn4.0Ag0.5Cu. The package surface finish was Ni/Au. The printed circuit board was made of FR-4 with a thickness of 2.34 mm (92 mil) and a size of 220 mm × 140 mm. The board finish was OSP. The stencil used was 0.127 mm (5 mil) thick with openings of 0.457 mm (18 mil) diameter. Two reflow profiles were used. One was 205°C peak temperature and 67 seconds over 183°C, and the other was 214°C peak temperature and 77 seconds over 183°C. The microstructures of the solder joints are shown in Fig. 7.5. It is clearly shown that Pb was partially diffused into the SnAgCu ball when the reflow peak temperature was at 205°C. A nearly full mixing was achieved when the reflow peak temperature was at 214°C, but the dispersion of Pb was not uniform.

Zbrzezny et al. investigated various reflow profiles and concluded that complete mixing of the solders was achieved when the reflow peak temperature reached 218 - 222°C [6].

Most of these studies believed that full mixing was achieved only when the reflow peak temperature exceeded 217°C [4, 6, 7], however, a full mixing of the SnPb paste with the SnAgCu ball can be achieved when the peak reflow temperature is below 217°C. For example, Nandagopal et al. observed that a full mixing of the SnPb paste and the SnAgCu ball was accomplished at a peak reflow temperature of 210°C for about 15 to 25 seconds [8, 9]. They used the Differential Scanning Calorimeter (DSC) to characterize the time required to achieve full mixing. Handwerker indicated that full mixing of the tin-lead paste and lead-free Sn3.9Ag0.6Cu solder ball

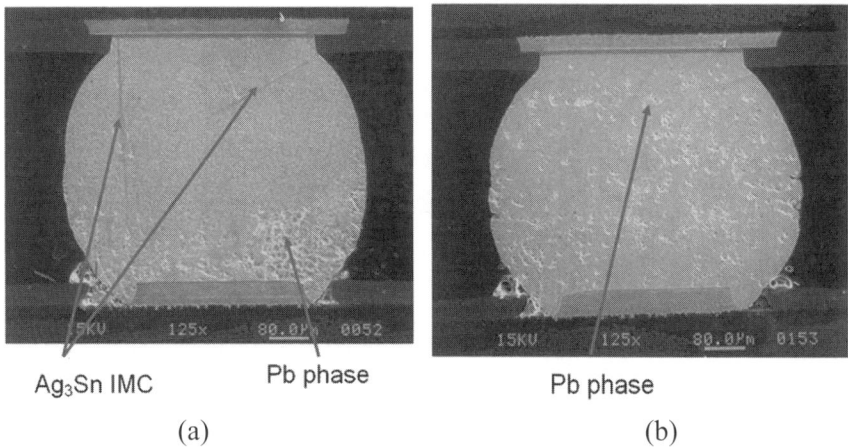

Ag₃Sn IMC   Pb phase   Pb phase

(a)   (b)

Fig. 7.5. SEM pictures of reflowed at a) 205°C peak temperature and 67 seconds over 183°C; b) 214°C peak temperature and 77 seconds over 183°C

occurred at 207°C with a sufficient time, the Sn3.9Ag0.6Cu solder ball constituting 75% of the final solder [10]. Snugovsky et al. described the mixing process using a SnPb phase diagram [11]. From the study, they concluded that a complete mixture may be achieved at a temperature lower than 217°C and that the temperature depends on solder ball composition, ball/solder paste ratio, dwell time, and component size.

7.2.2 Reliability of BGA/CSP Backward Compatibility

A significant number of experimental studies have been done recently on investigating the solder joint reliability of BGA/CSP backward compatibility using various reflow profiles [5, 6, 7, 8, 9, 11, 12, 13, 14, 15, 16].

Poor Reliability when the BGA/CSP Ball is Partially Mixed

It is evident that the reliability of solder joint interconnections in backward compatibility assemblies degrades significantly if SnAgCu solder spheres are only partly melted in backward compatibility. Hillman et al. evaluated the reliability of a BGA package assembled using a peak reflow temperature of 215°C with the duration time above 200°C at 40 seconds [12]. They observed partial mixing of Pb in the joint microstructure. The reliability of the solder joint was very poor as the solder joint failed at only 137 cycles in temperature cycling from –55°C to +125°C with dwell times of 11 minutes at each extreme and a ramp rate of 10°C/minute maximum per IPC-9701 guidelines. They used BGA/CSP components with Sn4.0Ag0.5Cu solder alloy. The reflow profile was developed using thermocouples attached to the outside edge of the BGA solder spheres with conductive epoxy. The test vehicle was FR4, 2.08 mm (82 mil) thick, and a glass transition temperature of 170°C minimum. The size of the board was 203 mm × 279 mm (8 inch × 11 inch) and the board finish was ENIG with 18 layers of ½ ounce Cu. The component under test was 256 I/O daisy chained, 17 mm ×17 mm, 1.0 mm pitch, full array, Sn4Ag0.5Cu solder ball alloy. The stencil used was 0.127 mm (5 mil) thick, with 1:1 board pad to stencil aperture match and 0.381 mm (15 mil) diameter round apertures. Though solder paste volume was not measured, 95% paste transfer ratio was assumed.

Gregorich & Holmes reported that the reliability of backward compatibility assemblies when the mixed assembly was reflowed at the peak temperature of 200°C and the duration of time above 183°C at 62 seconds was much poorer than that of the control SnAgCu ball with SnAgCu paste in both the accelerated temperature cycling test from –40°C to +125°C and the mechanical shock test [13]. The poor reliability was believed to be due to the inhomogeneous microstructure resulting from partial mixing of Pb. The reliability of backward compatibility assembly improved as the reflow temperature increased to 225°C. The package investigated was a CSP with 0.5 mm pitch. These components were 14 mm × 14 mm × 1.2 mm in size. The solder spheres in the component were Sn4.0Ag0.5Cu. The PCB was 4-layer, 0.8 mm thick FR-4 board, 100 mm × 40 mm in size, with Ni/Au board surface finish. The stencil used in the study was 0.127 mm (5 mil) thick and 1:1 ratio between package land and board land.

Hua et al. reported similar results showing that incomplete mixing leads to unacceptable solder joints [7, 14]. Therefore, it is critical to achieve complete mixing of SnPb paste with SnAgCu ball in BGA/CSP backward compatibility assembly.

Reliability Comparison between Backward Compatibility versus SnPb Control and SnAgCu Control Assemblies

If complete mixing is achieved in backward compatibility assemblies, is the reliability better, equivalent or worse than that of SnPb control and SnAgCu control assemblies? There are conflicting reports about whether the reliability of backward compatibility assembly is better or poorer than that of SnAgCu balls assembled with SnAgCu solder paste. In general it is equivalent or worse. Bath et al. found that the reliability of backward compatibility assembly in accelerated temperature cycling (ATC) from 0°C to 100°C with 40 minute a cycle, even when the full mixing was achieved, was poorer than that of both SnAgCu ball/SnAgCu paste and SnPb ball/ SnPb paste as shown in Fig. 7.6 [5]. It should be pointed out that there is no statistically significant difference in reliability between the reflow peak temperature of 205°C and 215°C. But Bandagopal et al. found that the reliability of backward compatibility assembly in both ATC from 0°C to 100°C and –40°C to 125°C was better than the SnPb assembly when full mixing was achieved [8]. Bandagopal et al. also found that the reliability of backward compatibility assemblies surpassed the reliability of SnAgCu control assemblies in ATC from –40°C to 125°C, but not in ATC from 0°C to 100°C.

Although a considerable amount of work has been done so far on the backward compatibility assembly and its reliability, the minimum temperature able to achieve full mixing is still unknown. The key in backward compatibility assembly is to develop a reflow profile with the peak temperature high enough to be able to achieve full mixing of the SnPb paste and the SnAgCu ball, but low enough (prefer below 220°C) so that SnPb components and the board won't be damaged. Therefore, it is critical to know the minimum reflow peak temperature that is capable of achieving a complete mixing of SnPb paste with lead-free components.

Work in the iNEMI backward compatibility group is helping to define and understand the peak temperature and time above liquidus to use for certain types of lead-free CSP/BGA components with tin-lead paste which

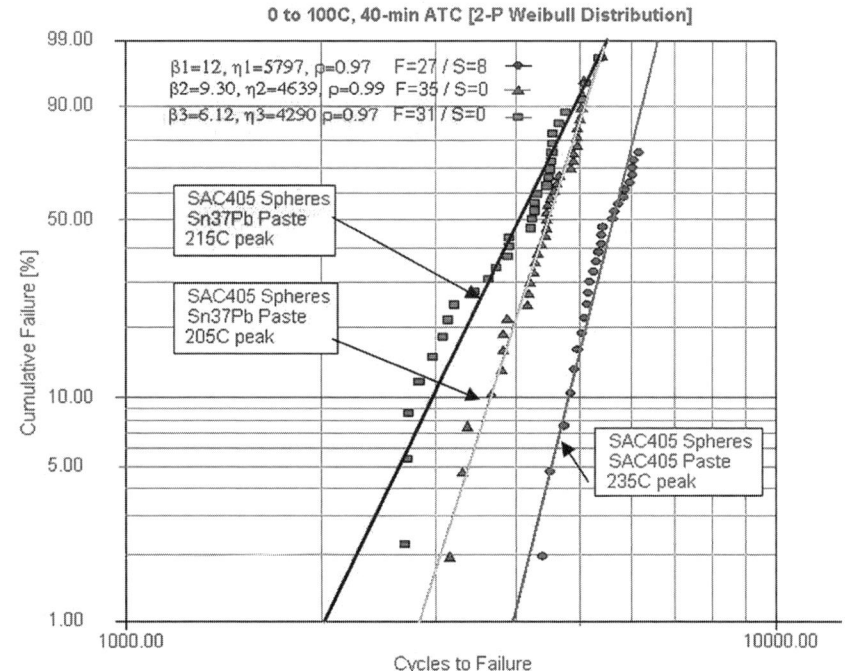

Fig. 7.6. Backward reliability data [5]

will be put into ATC reliability testing [17]. Components to be ATC reliability tested in iNEMI project will be the SBGA600 package with a pitch of 1.27 mm and a size of 45 mm × 45 mm, the PBGA324 package with a pitch of 1 mm and a size of 23 mm × 23 mm, the CABGA288 package with a pitch of 0.8 mm and a size of 19 mm × 19 mm, and the CTBGA132 package with a pitch of 0.5 mm and a size of 8 mm × 8 mm.

7.3 Estimation of Mixed Composition Liquidus Temperature

7.3.1 Mixed Composition Calculation

There are four alloying elements in the mixed composition when SnAgCu BGA/CSP components are soldered with SnPb paste: Sn, Ag, Cu, and Pb. The percentage of each metal element in the mixed composition can be calculated [18, 19]

$$W_{Pb} = \frac{f_{Pb} \times V_{Paste} \times f_m \times d_{SnPb}}{V_{Paste} \times f_m \times d_{SnPb} + V_{Ball} \times d_{SnAgCu}} \quad (7.1)$$

$$W_{Ag} = \frac{f_{Ag} \times V_{Ball} \times d_{SnAgCu}}{V_{Paste} \times f_m \times d_{SnPb} + V_{Ball} \times d_{SnAgCu}} \quad (7.2)$$

$$W_{Cu} = \frac{f_{Cu} \times V_{Ball} \times d_{SnAgCu}}{V_{Paste} \times f_m \times d_{SnPb} + V_{Ball} \times d_{SnAgCu}} \quad (7.3)$$

$$W_{Sn} = 100 - W_{Pb} - W_{Ag} - W_{Cu} \quad (7.4)$$

where W_{Pb}, W_{Ag}, W_{Cu}, and W_{Sn} are the weight percentages of Pb, Ag, Cu, and Sn in the mixed compositions, respectively; f_{Pb} is the percentage of Pb by weight in SnPb solder paste; f_{Ag} and f_{Cu} are the weight percentage of Ag and Cu in SnAgCu alloy; f_m is the volume percentage of metal content in SnPb solder paste; d_{SnPb} and d_{SnAgCu} are the density of SnPb and SnAgCu alloys. V_{paste} is the SnPb solder paste volume, which can be calculated

$$V_{paste} = \begin{cases} L^2 H(TR) & \text{for square aperature} \\ \pi \left(\dfrac{D}{2}\right)^2 H(TR) & \text{for round aperture} \end{cases} \quad (7.5)$$

where L is stencil aperture length for square aperture, H is stencil thickness, D is stencil aperture diameter for round aperture, and TR is the paste transfer ratio, which is defined as the ratio of the volume of solder paste deposited to the volume of the aperture.

V_{ball} is the volume of a solder ball in the BGA/CSP component. If the ball diameter, D, is given, the ball volume can be calculated

$$V_{ball} = \frac{4}{3}\pi\left(\frac{D}{2}\right)^3 \tag{7.6}$$

If the sphere is reflowed and the ball height, H, and radius, R, are given, the ball volume can be calculated

$$V_{ball} = \frac{2}{3}\pi R^3 - \pi R^3\left[\frac{1}{3}\left(\frac{H-R}{R}\right)^3 - \left(\frac{H-R}{R}\right)\right] \tag{7.7}$$

For eutectic SnPb solder paste, f_{Pb} is 37 and typical value of f_m is 0.5 (or 50%). For Sn3.0Ag0.5Cu solder alloy, f_{Ag}=3.0 and f_{Cu}=0.5. The density of eutectic Sn37Pb, d_{SnPb}, is 8.4 g/cm^3 and the density of Sn4.0Ag0.5Cu, d_{SnAgCu}, is 7.394 g/cm^3 [20].

Use the Bath et al. study [5] as an example. A 1 mm pitch BGA (FG676) package used with a 0.61 mm (24 mil) Sn3.0Ag0.5Cu ball diameter was assembled with Sn37Pb paste. The solder paste was printed using a 0.127 mm (5 mil) thick stencil with a 0.457 mm (18 mil) diameter round stencil aperture. The SnPb paste had 50% metal content in volume. Assuming a 90% solder paste transfer ratio, using Eq. 7.6, we can calculate that the volume of the Sn3Ag0.5Cu ball is 0.118 mm^3 (7235 mil^3). Using Eq. 7.5, we can calculate the volume of SnPb paste as 0.0188 mm^3 (1125 mil^3). Using Eqs. 7.1 to 7.4, we can get the final mixed alloy composition:

$$W_{Pb} = \frac{37 \times 0.0188 \times 0.5 \times 8.4}{0.0188 \times 0.5 \times 8.4 + 0.118 \times 7.394} = 3.07$$

$$W_{Ag} = \frac{3.0 \times 0.118 \times 7.394}{0.0188 \times 0.5 \times 8.4 + 0.118 \times 7.394} = 2.75$$

$$W_{Cu} = \frac{0.5 \times 0.118 \times 7.394}{0.0188 \times 0.5 \times 8.4 + 0.118 \times 7.394} = 0.46$$

$$W_{Sn} = 100 - 3.07 - 2.75 - 0.46 = 93.7$$

Therefore, the mixed alloy joint composition is 93.7% Sn, 3.1% Pb, 2.8% Ag, and 0.5% Cu, all in weight.

7.3.2 Estimation of Mixed Composition Liquidus Temperature

After we know the mixed compositions, the next question becomes what in this case is the liquidus temperature of the mixed composition Sn3.-1Pb2.8Ag0.5Cu. The phase diagram of common binary and ternary systems that are relevant to solders is available at the Fundamental Properties of Pb-Free Solder Alloys Chapter in this book. But the phase diagram of the complex quaternary SnPbAgCu is currently not available. The phase equilibria can be calculated from thermodynamic databases using the CALPHAD method [21]. Thermodynamic calculation is a very useful tool in obtaining phase diagram information, but it requires reliable thermodynamic databases and specialized knowledge.

Kattner and Handwerker stated that the liquidus temperature of ternary and quaternary systems could be calculated using the simple linearization of the binary liquidus lines [22]

$$T_l = 232°C - 3.1W_{Ag} - 1.6W_{Bi} - 7.9W_{Cu} - 3.5W_{Ga} \qquad (7.8)$$
$$-1.9W_{In} - 1.3W_{Pb} + 2.7W_{Sb} - 5.5W_{Zn}$$

limits: $W_{Ag} < 3.5$; $W_{Bi} < 43$; $W_{Cu} < 0.7$; $W_{Ga} < 20$; $W_{In} < 25$; $W_{Pb} < 38$; 38; $W_{Sb} < 6$; $W_{Zn} < 6$

where T_l is the liquidus temperature of Sn-rich solder alloys, 232C is the liquidus temperature of Sn, W_{Ag}, W_{Bi}, W_{Cu}, W_{Ga}, W_{In}, W_{Pb}, W_{Sb}, W_{Zn} is the percentage in weight of Ag, Bi, Cu, Ga, In, Pb, Sb, and Zn, respectively. The coefficient before these alloying elements is the slope of the binary liquidus lines. For example, 7.9 is the slope of SnCu binary liquidus lines when Cu is less than 0.7% in weight; 1.3 is the slope of SnPb binary liquidus lines when Pb is less than 38% in weight; and so on. It should be emphasized that the limitation of the simple linearization is $W_{Ag} < 3.5$; $W_{Bi} < 43$; $W_{Cu} < 0.7$; $W_{Ga} < 20$; $W_{In} < 25$; $W_{Pb} < 38$; $W_{Sb} < 6$; $W_{Zn} < 6$. It should also be noted that Eq. 7.8 is an approximation.

Based on Eq. 7.8, the liquidus temperature of the quaternary SnPbAgCu system, a typical alloy system in both forward compatibility assembly and backward compatibility assembly, can be calculated

$$T_l = 232°C - 3.1W_{Ag} - 7.9W_{Cu} - 1.3W_{Pb} \qquad (7.9)$$

with limits: $W_{Ag} < 3.5$; $W_{Cu} < 0.7$; $W_{Pb} < 38$;

Based on Eq. 7.9, the liquidus temperature of Sn3.0Ag0.5Cu is,

$$T_l = 232°C - 3.1 \times 3.0 - 7.9 \times 0.5 - 1.3 \times 0 = 219°C$$

If Ag content is over 3.5% and less than 4% wt, Ag$_3$Sn is primary phase. In this case, Eq. 7.9 is not valid. A simple fix is to add 5°C to Eq. 7.8. Thus, the liquidus temperature of Sn3.8Ag0.7Cu is

$$T_l = 232°C - 3.1 \times 3.8 - 7.9 \times 0.7 - 1.3 \times 0 + 5 = 220°C$$

Currently the reflow profiles in backward compatibility assembly are developed through costly trial-and-error methods. It is expected that the estimation of the mixed composition liquidus temperatures will be able to guide process engineers to develop the right reflow profile in backward compatibility assembly.

Table 7.4 summarizes the final joint compositions and liquidus temperature with SnAgCu ball and Sn37Pb paste for typical BGA/CSP component pitch levels. The aperture size, shape, stencil thickness and ball diameter are based on typical guidelines for no-clean paste. The transfer ratio is assumed based on experience. It shows that the final liquidus temperature is lower than 217°C, the liquidus temperature of SnAgCu. The liquidus temperature can be as low as 201°C. As the component pitch decreases (except for the case of 0.5 mm pitch), the weight percentage of Pb increases and the liquidus temperature decreases. Eqs. 7.1 to 7.4 imply that the liquidus temperature depends on the ratio of BGA ball volume and solder paste volume.

Table 7.4. Final Joint Compositions and Liquidus Temperature with SnAgCu Ball and Sn37Pb Paste

Pitch (mm)		1.27	1.0	0.8	0.65	0.5
Aperture size in mm		0.533	0.457	0.406	0.356	0.279
(mil)		(21)	(18)	(16)	(14)	(11)
Aperture shape		Round	Square	Square	Square	Square
Stencil thickness in mm		0.152	0.127	0.127	0.127	0.102
(mil)		(6)	(5)	(5)	(5)	(4)
Solder paste transfer ratio (%)		100	90	85	80	90
Ball diameter in mm		0.711	0.559	0.356	0.254	0.254
(mil)		(28)	(22)	(14)	(10)	(10)
Weight % of Pb		3.4	4.8	11.1	17.0	11.9
	Weight % of Ag	2.7	2.6	2.1	1.6	2.1
SAC305	Weight % of Cu	0.5	0.4	0.3	0.3	0.3
Ball	Estimated liquidus temperature (°C)	216	214	208	203	208
	Weight % of Ag	3.6	3.5	2.8	2.2	2.7
SAC405	Weight % of Cu	0.5	0.4	0.3	0.3	0.3
Ball	Estimated liquidus temperature (°C)	218	217	206	201	205

7.3.3 Effect of Pb Content on Backward Compatibility Reliability

Eq. 7.9 shows that a higher Pb percentage in the mixed composition can reduce the mixed composition liquidus temperature. The higher Pb percentage can be achieved by printing more SnPb solder paste or reducing the SnAgCu solder ball volume. But increasing the lead content in the solder joint can also lead to more issues as lead tends to segregate at the tin grain boundaries which can be a source of crack initiation or propagation. Zhu et al. studied the effect of Pb contamination on the lead-free solder joint microstructure and observed a Pb-rich phase formed in the bulk solder when the lead-free solder contains Pb impurity [23]. Zeng discussed the influence of the Pb-rich phase on solder joint reliability [24]. The Pb-rich phase may be the weakest region in the bulk solder, and the crack may propagate along the Pb-rich phase interface during reliability testing.

However, some experimental results did not follow the explanation. Bandagopal et al. found that the reliability of backward compatibility assembly in both ATC from 0°C to 100°C and –40°C to 125°C was better than the SnPb assembly when full mixing was achieved [9]. Furthermore, the reliability data of SnPb BGA ball soldered with SnAgCu paste (or forward compatibility), where higher Pb content existed in the mixed compositions, was better or equal to that of SnPb ball/SnPb paste control assemblies [25]. Hunt and Wickham concluded that there should be few solder joint reliability problems when mixing SnPb and lead-free components and solder alloys (with lead contamination in the range of 1 to 10%) [26]. Therefore, it is difficult to draw a conclusion regarding the effect of Pb content on backward compatibility reliability although there is evidence to show that it could be detrimental [12]. For the case where the lead-free SnAgCu paste is assembled with SnPb BGA/CSP components, if the voiding is excessive, this may lead to reliability issues from excess voiding which reduce the effective solder cross-sectional area or if the ball size and pitch is small, bridging may occur between adjacent spheres.

7.3.4 Comparison of Estimated Liquidus Temperature and the Experimental Results

To assess the method to calculate estimated liquidus temperatures, calculated liquidus temperatures were compared with published experimental results. The estimated temperatures and the published experimental results are summarized in Table 7.5. If the reflow peak temperature used was higher than the estimated liquidus temperature, full mixing was expected.

Table 7.5. Comparison of the Estimated Liquidus Temperature and the Reported Experimental Results

Reference	Estimated liquidus temperature	Peak reflow temperature used	Experimental results
Gregorich & Holmes [13]	209°C	200°C	Partial mixing
		225°C	Full mixing
Hillman et al. [12]	219°C	215°C	Partial mixing
Grossmann et al. [4]	216°C	210°C	Partial mixing
		217°C	Full mixing
Nandagopal et al. [9]	212°C	210°C	Full mixing
		227°C	Full mixing
Bath et al. [5]	218°C	205°C	Partial mixing
		214°C	Full mixing

Otherwise, partial mixing was expected. Overall, Table 7.5 shows that the estimated liquidus temperatures are generally consistent with reported experimental results. There are small variances between the estimated temperature and the reported results of studies in Nandagopal et al. [9] and Bath et al. [5]. This could be due to the inaccuracy of paste transfer ratio assumptions and the fact that a sufficient time over liquidus (183°C) for the tin-lead solder paste could also affect the result. Since only a few studies have reflow peak temperatures close to the estimated liquidus temperature, further experimental study is needed to validate the accuracy of the estimation method.

7.4 Chip Component and Lead-Frame Component Backward Compatibility

7.4.1 Calculation of Mixture Composition Liquidus Temperature for Chip Terminations

The typical surface finish for SnPb chip terminations is 90Sn10Pb with a liquidus temperature of 219°C. The typical surface finish for lead-free chip terminations is 100% Sn with a liquidus temperature of 232°C. Using the 0603 chip as an example, given chip component dimensions [27, 28] and a final plating of 100% Sn (lead-free) or 90% Sn10% Pb (SnPb) of 7.5 to 15 micron thickness, the solder volume in the chip component can be calculated. Knowing the component size and typical stencil apertures, the SnPb solder paste volume deposited can be calculated.

Table 7.6 summarizes the calculated minimum and maximum liquidus temperature of mixed compositions with Sn37Pb paste based on the minimum

Table 7.6. Calculated Final Liquidus Temperature of Chip Component Solder Joint

Chip Compo- nent Size	2512	1206	0805	0603	0402	0201
Lead-free (100% Sn)	187 – 190°C	189 – 192°C	188 – 191°C	191 – 196°C	198 – 206°C	195 – 202°C
SnPb (90Sn10Pb)	186 – 189°C	187 – 190°C	187 – 189°C	189 – 193°C	195 – 200°C	192 – 197°C

(7.5 micron) and maximum (15 micron) coating thickness mentioned. It shows that there is no significant increase in mixed composition liquidus temperature from 90Sn10Pb to a lead-free pure tin termination. Another point to make is that the final alloy composition for both SnPb and lead-free components are similar, which is close to Sn37Pb. The reason is that the coating thickness on a chip component is considerably thinner than that for a BGA/CSP component so it does not significantly affect the mixed solder joint alloy composition.

7.4.2 Calculation of the Solder Joint Mixed Composition Liquidus Temperature for Lead-Frame Components

The typical surface finish for SnPb lead-frame components is 90Sn10Pb with a liquidus temperature of 219°C. There are several common surface finishes for lead-free leadframe components, for example, 100% Sn, Sn3.5Ag, Sn1.0Cu, Sn2-4% wt. Bi, NiPdAu and NiAu. Different surface finishes have their own advantages and disadvantages. In this chapter, the liquidus temperature calculation is limited to the most common lead-frame surface finish, 100% Sn.

The typical plating thickness for leadframe components of 100% Sn (lead-free) or 90% Sn10% Pb (SnPb) is 7.5 to 15 micron. Given a leadframe component dimensions [27, 28], the solder volume in the leadframe component can be calculated. Knowing the component pitch and its appropriate stencil aperture openings and paste transfer ratios, the solder paste volume can be calculated.

Table 7.7 summarizes the calculated minimum and maximum liquidus temperature of mixed component joint compositions with Sn37Pb paste based on the minimum (7.5 micron) and maximum (15 micron) coating thickness mentioned for leadframe components. It shows that there is no significant increase in the mixed joint composition liquidus temperature from 90Sn10Pb termination to the lead-free pure tin termination similar to the case of the chip component. Again the final alloy composition for both SnPb and lead-free pure tin components with SnPb paste are similar, which is close to the composition of Sn37Pb solder.

Table 7.7. Calculated Final Liquidus Temperature of Leadframe Component Solder Joint

Component Lead-frame Pitch (mm)	1.0	0.8	0.65	0.5	0.4
Lead-free (100% Sn)	191 – 197°C	188 – 192°C	187 – 190°C	187 – 189°C	188 – 196°C
SnPb (90Sn10Pb)	190 – 194°C	187 – 190°C	186 – 189°C	186 – 188°C	187 – 189°C

7.4.3 Backward Compatible Solder Joint Reliability of Chip and Lead-frame Components

The reliability of chip and lead-frame component solder joints in backward compatibility assemblies is not expected to be significantly different from the SnPb control assemblies. Since the volume of SnPb solder paste is significantly greater than that of the solder in surface finish of the chip or lead-frame components, the final alloy composition in backward compatibility is similar to Sn37Pb solder. Thus the liquidus temperature and the reflow profile needed for the final mixed compositions are similar to the eutectic SnPb solder. There are many years of historical data showing that lead-free chip and lead-frame components in backward compatibility assemblies are reliable. Table 7.8 lists the reliability data of the 2.36 mm (93 mil) Solectron lead-free surface mount test board after ATC from 0 to 100°C for 6,013 cycles. It shows that the reliability of backward compatibility assemblies is comparable to that of SnPb control assemblies.

Table 7.8. Reliability Data of Lead-free Pure Sn Leadframe Coatings with SnPb Paste

	Reflow peak temperature	Number of samples failed over number of samples tested		First failure occurred (cycles)	
		Sn10Pb	Leadfree pure Sn	Sn10Pb	Leadfree pure Sn
28mm QFP256 0.4mm pitch	205 - 215°C	2/32	2/16	1,022	2,213
SOIC20 1.27mm pitch	205 - 215°C	0/32	0/16	–	–
PLCC20 1.27mm pitch	205 - 215°C	1/32	2/16	3,595	1,305
7mm MLF 0.5mm pitch	205 - 215°C	10/48	6/24	603	4,070

In Japan, OEMs have successfully used lead-free SnBi component coatings for lead-free product. The question that has created concern is the use of lead-free SnBi component coatings with SnPb soldering materials due to the potential for the formation of low melting point phases. Work has been done by NIST (National Institute of Science and Technology) to understand the SnBiPb phase diagram [29]. The ternary eutectic composition is approximately 51.5Bi15.5Sn33Pb with a melting temperature of 96°C. Considering a thickness of 0.076 mm (3 mil) reflowed 63Sn37Pb solder joint and a 97Sn3Bi component coating with a coating thickness of 10 micron, the final composition of the joint is around 62.7Sn37Pb0.3Bi (< 1wt% Bismuth). Based on calculations by NIST the solidus temperature of 62.7Sn37Pb0.3Bi is 174°C so the 96°C ternary eutectic would not form. Calculations have also been done for different SnPbBi compositions namely Sn37Pb1Bi with a solidus temperature of 159°C and Sn37Pb3Bi with a solidus temperature 119°C.

The JEDEC/IPC JP002 document [30] indicates that the ternary eutectic phase will not form using Sn1-4wt%Bi coatings with Sn37Pb solder. For most components SnBi plating is acceptable for use with SnPb solder. But there may be a risk of excessive intermetallic growth if the storage product temperature exceeds 135°C. Excessive IMC was observed in ageing experiments with SnAgCu soldered SnBi coated surface mount components [31]. Additional reliability testing is needed to validate its use at elevated storage temperatures.

For tin-lead wave soldering, there are still potential restrictions on the use of SnBi components because the bismuth can leach into a tin-lead wave solder pot and bismuth could accumulate over time in the solder pot, which can lead to a lowering of the melting temperature of the pot and potential reliability issues such as fillet lifting.

7.5 Forward Compatibility

Only a few studies have been published on the reliability of forward compatibility assemblies, or lead-free solder paste (SnAgCu) assembled with tin-lead BGA/CSP and lead-frame and chip components. Though more voids were observed in forward compatibility assemblies for BGAs/CSPs [1], the reliability of forward compatibility assemblies is equivalent or better than the reliability of the SnPb balled BGAs with SnPb solder paste [2, 25, 32].

Nurmi and Ristolainen reported that forward compatibility assemblies did not show any serious reliability risks and can withstand temperature cycling stress better than SnAgCu control assemblies [25]. Experimental studies from the iNEMI lead-free assembly project found no ATC reliability problems in forward compatibility assemblies as well as shown in Table 7.9 [2]. This included lead-free SnAgCu paste with tin-lead CSP169, CSP208, PBGA256, CBGA256 and 2512 chip resistors. Lau et al. concluded that the quality of the SnPb balled FLEXBGA solder joints with lead-free solder paste on Ni-Au PCB is better than that with SnPb solder paste on an OSP PCB with 99 percent confidence level [32]. Note that all studies in forward compatibility assembly used SnAgCu reflow profiles.

However, Seelig et al. found that a lead-free SnAgCu soldered joint with a tin-lead coated leadframe component caused a concentration of lead at the joint/board surface interface which was the last area of the solder joint to solidify [33]. The SnPbAg with a melting temperature of 179°C could be present in this area. The resulting solder joint was found to have a weak interface at this point.

For tin-lead coated through-hole components or tin-lead HASL coated boards waved with lead-free solder, the risk of lead contamination of a lead-free wave solder pot (>0.1wt% lead which is the European Union ROHS limit) would mean that this specific mixing should not be attempted. Fillet lifting may also occur in the wave soldered joint.

There are issues as to whether the tin-lead component or other components on the board or the board itself are rated to the higher lead-free soldering temperatures. Based on our knowledge, this has not been discussed in published literature. In addition, the resulting solder joint would not be compliant to legislative direction.

Table 7.9. Relative ATC Performance [2]

Component (ImAg board finish unless indicated)	−40°C to 125°C			0°C to 100°C		
	Pb	Mixed	LF	Pb	Mixed	LF
48 TSOP	0	–	0			
48 TSOP, NiAu boards	0	+	+			
R2512 resistor	0	0	0			
R2512, NiAu boards	0					
169CSP	0	+	+	0	0	+
208CSP	0	0	+	0	+	+
208CSP, JEITA alloy (Sn3Ag0.5Cu)			0			
256PBGA	0	0	0	0	0	0
256CBGA				0	–	+

Note: 0 = equivalent, + = superior, − = inferior

7.6 Press Fit Connector Interconnections

7.6.1 Introduction for Press-Fit Connectors

The European Union RoHS Directive covers various aspects of electronics manufacturing including press-fit interconnections. In spite of the current exemption for lead used on compliant pin connector systems, the transition to lead-free press-fit interconnect components is inevitable and in progress. OEMs, EMSs, and connector suppliers are working together to make the connectors used in electronics products RoHS compliant. In the transition period, lead-free press-fit connectors are required to have backward and forward compatibility meaning they can be used in both lead-free and tin-lead assembly processes on any surface finish selected.

7.6.2 Current Status of Lead-Free Press-Fit Connectors

In the area of press-fit interconnections, lead-free impacts are seen in the changes of press-fit connector compliant pin plating, PCB laminate material, and board surface finish. For press-fit connector compliant pins, the main plating includes matte Sn and electroplated Au (over nickel). For the PCB, the major board surface finishes include Immersion Sn, ENIG (NiAu), Immersion Ag, and OSP. The combination of the compliant pin plating with the board surface finish will generate various press-fit insertion force results as opposed to tin-lead connector compliant pin plating with the board surface finish. To date, no comprehensive studies and industry-accepted conclusions are available that satisfy all design and assembly conditions. Nevertheless, there are a few mixed observations obtained from production experience or derived from OEM/EMS/connector supplier studies:

- Lead-free compliant pins experienced increased insertion forces due to interactions among various pin design attributes (eye-of-needle and other custom-made compliant sections), compliant pin plating type, and board surface finish type, as compared to its tin-lead coated counterpart compliant pins given the same PCB Plated Through Hole (PTH) size. Failures such as bent pins and PTH damage have been seen when pressing in lead-free coated compliant pins into the board [34].
- Depending on the compliant pin design, PCB surface finish type, the pin-to-hole area ratio, the lead-free coated compliant pin can have as hihg as a 15% average increase in insertion force over tin-lead coated compliant pin [35].
- According to a study [36], the NiAu board surface finish caused higher insertion/retention force as opposed to other board surface

finishes tested. This study recommended immersion Ag as a suitable choice among RoHS compliant PCB finishes in terms of relatively lower insertion/retention force with lead-free press-fit connectors.

- Another study showed that immersion Sn caused highest insertion forces with the lead-free press-fit connector, among six board surface finishes tested [37].
- The actual PCB laminate material used (such as Phenolic non-dicy laminate) that is rated for lead-free reflow conditions might contribute partially to the increase of insertion forces with lead-free press-fit connectors as it may be harder and less forgiving than the standard laminate material (dicy laminate) used for typical tin-lead soldering. This phenomenon is yet to be fully understood and studied.

With the increase of compliant pin insertion forces, changes in design and assembly processes have to take place accordingly to ensure that the yield and performance of lead-free press-fit interconnection are not compromised. Long-term solutions for the lead-free press-fit connector supplier could be 1) to change the compliant pin geometry design to be more suitable for lead-free insertion assembly, or 2) to study the lead-free compliant connectors' press-fit behavior on various lead-free PCB surface finishes and update their connector specifications accordingly. Before this can materialize, short-term solutions include 1) reducing the connector pressing speed into the board, 2) using lubricant to reduce insertion force, or 3) changing PCB PTH drill and/or finished hole size.

7.6.3 Lead-Free Press-Fit Connector Summary

Due to the press-fit compliant pin plating changing from SnPb to a lead-free finish, the press/insertion process needs to be re-characterized. Currently there are only limited studies done by OEM/EMS/connector suppliers in this regard. No conclusion can be drawn yet in terms of the selection of the best PCB surface finish, PCB laminate material, and compliant pin plating. It is recommended that EMS providers and their OEM customers, as well as connector suppliers work together to make the lead-free press-fit interconnection transition smooth without compromising quality and reliability.

Summary

In this chapter, backward compatibility and forward compatibility have been reviewed with emphasis on the reliability of BGA/CSP backward compatibility assemblies. It is evident that the reliability of solder joint interconnections in backward compatibility assemblies degrades significantly if SnAgCu solder spheres are only partly melted in backward compatibility. If complete mixing is achieved in backward compatibility assemblies, there are conflicting experimental results on the reliability of BGA/CSP backward compatibility. Data show that the backward compatibility assemblies of chip components and leadframe components are reliable in terms of solder joint integrity.

The estimation of the liquidus temperature of mixed composition in backward compatibility has been presented for BGA/CSP, lead-frame and chip components. The estimation for BGA/CSP components could be used to guide the development of a reflow profile, but it should be noted that the estimation is an approximation and further experimental study is needed to validate the accuracy of the method.

The majority of forward compatibility studies show little or no issues but excessive voiding of tin-lead BGA/CSP components is a concern. The effect of Pb content in mixed assemblies (forward compatibility and backward compatibility) is still questionable.

Both the backward and forward compatibility situations should be considered as transitional processes only with a full movement to lead-free paste with lead-free components being the general goal to avoid any reliability issues associated with the two transition assembly situations.

Acknowledgements

The first author, J. Pan would like to express his gratitude to the Solectron Corporation, Milpitas, CA, for technical and financial support; to Kim Hyland, Dennis Willie, without whom this work would not have been possible. He would also like to thank Dr. Ursula Kattner and Dr. Carol Handwerker for their valuable technical discussions.

References

1. Smetana J, Horsley R, Lau J, Snowdon K, Shangguan D, Gleason J, Memis I, Love D, Dauksher W, Sullivan B (2004) Design, materials and process for lead-free assembly of high-density packages. Soldering & Surface Mount Technology, 16 (1): 53-62.
2. Handwerker C, Bath J, Benedetto E, Bradley E, Gedney R, Siewert T, Snugovsky P, Sohn J (2003) NEMI Lead-free Assembly Project: Comparison Between PbSn and SnAgCu Reliability and Microstructures. In: Proceedings of the SMTA International Conference.

3. Lau JH, Liu K (2004) Global trends of lead-free soldering. Advanced Packaging Jan. 2004.
4. Grossmann G, Tharian J, Jud P, Sennhauser U (2005) Microstructural investigation of lead-free BGAs soldered with tin-lead solder. Soldering & Surface Mount Technology 17 (2): 10-21.
5. Bath J, Sethuraman S, Zhou X, Willie D, Hyland K, Newman K, Hu L, Love D, Reynolds H, Kochi K, Chiang D, Chin V, Teng S, Ahmed M, Henshall G, Schroeder V, Nguyen Q, Maheswari A, Lee MJ, Clech J-P, Cannis J, Lau J, Gibson C (2005) Reliability Evaluation of Lead-free SnAgCu PBGA676 Components using Tin-Lead and Lead-free SnAgCu solder paste. In: Proceedings of 2005 SMTA International, Chicago, IL, 891-901.
6. Zbrzezny AR, Snugovsky P, Lindsay T, Lau R (2005) Reliability Investigation of Sn-Ag-Cu BGA Memory Modules Assembled with Sn-Pb Eutectic Paste Using Different Reflow Profiles. In: International Conference on Lead-free Soldering, CMAP, Toronto, Ontario, Canada, May 24-26, 2005.
7. Hua F, Aspandiar R, Anderson C, Clemons G, Chung C, Faizul M (2003a) Solder Joint Reliability Assessment of Sn-Ag-Cu BGA Components Attached with Eutectic Pb-Sn Solder. In: SMTA International 2005, Chicago, IL, pp. 246-252.
8. Nandagopal B, Chiang D, Teng S, Thune P, Anderson L, Jay R, Bath J (2005) Study on Assembly, Rework, Microstructures and Mechanical Strength of Backward Compatible Assembly. In: Proceedings of 2005 SMTA International, Chicago, IL, pp. 861-870.
9. Nandagopal B, Mei Z, Teng S (2006) Microstructure and Thermal Fatigue Life of BGAs with Eutectic Sn-Ag-Cu Balls Assembled at 210°C with Eutectic Sn-Pb Solder Paste. In: Proceedings of 2006 IEEE Electronic Components and Technology Conference, San Diego, CA, pp. 875-883.
10. Handwerker, C. (2005) Transitioning to Lead-free Assemblies, Printed Circuit Design and Manufacture, March Issue, pp. 17-23.
11. Snugovsky P, Zbrzezny AR, Kelly M, Romansky M (2005) Theory and Practice of Lead-free BGA Assembly using Sn-Pb Solder. In: International Conference on Lead-free Soldering, CMAP, Toronto, Ontario, Canada, May 24-26, 2005.
12. Hillman D, Wells M, Cho K (2005) The Impact of Reflowing a Pb free Solder Alloy Using a Tin/Lead Solder Alloy Reflow Profile on Solder Joint Integrity. In: International Conference on Lead-free Soldering, CMAP, Toronto, Ontario, Canada, May 24-26, 2005.
13. Gregorich T, Holmes P (2003) Low-Temperature, High Reliability Assembly of Lead-free CSPS. In: IPC/JEDEC 4th International Conference on Lead-free Electronic Components and Assemblies, Frankfurt, Germany, 2003.
14. Hua F, Aspandiar R, Rothman T, Anderson C, Clemons G, Klier M (2003b) Solder Joint Reliability of Sn-Ag-Cu BGA Components Attached with Eutectic Pb-Sn Solder Pate. Journal of Surface Mount Technology 16 (1): 34-42.
15. Sun F (2005) Solder Joint Reliability of Sn-Ag-Cu BGA and Sn-Pb Solder Paste. In: Proceedings of the 6th IEEE International Conference on Electronic Packaging Technology, Shenzhen, China.

16. Theuss H, Kilger T, Ort T (2003) Solder Joint Reliability of Lead-Free Solder Balls Assembled with SnPb Solder Paste. In: Proceedings of 2003 IEEE Electronic Components and Technology Conference, New Orleans, LA, pp. 331-337.
17. iNEMI (2006) Backward Compatibility Project Presentation. In: IPC Printed Circuits Expo, APEX and the Designers Summit, NEMI forum, Anaheim, CA, Feb. 8-10.
18. Pan J (2006) Estimation of Liquidus Temperature when SnAgCu BGA/CSP Components are Soldered with SnPb Paste. In: Proceedings of the 7th International Conference on Electronics Packaging Technology, Shanghai, China.
19. Pan J, Bath J (2006) Lead Free Soldering Backward Compatibility. In: IPC/JEDEC 12th International Conference on Lead Free Electronic Components and Assemblies, Santa Clara, CA, March 7-9, 2006.
20. Siewert T, Liu S, Smith DR, Madeni JC (2002) Database for solder properties with emphasis on new lead-free solders. NIST & Colorado School of Mines, Release 4.0, Feb. 2002, http://www.boulder.nist.gov/div853/lead free/solders.html
21. Kattner UR, Handwerker C (2001) Calculation of Phase Equilibria in Candidate Solder Alloys. Z. Metallkd. 92: 1-10.
22. Kattner UR, Handwerker C (2005) personal communication though emails.
23. Zhu Q, Sheng M, Luo L (2000) The effect of Pb contamination on the microstructure and mechanical properties of SnAg/Cu and SnSb/Cu solder joints in SMT. Soldering and Surface Mount Technology, 12 (3): 19-23.
24. Zeng X (2005) Thermodynamic analysis of influence of Pb contamination on Pb-free solder joint reliability. Journal of Alloys and Compounds, 348: 184-188.
25. Nurmi ST, Ristolainen (2002) Reliability of Tin-lead Balled BGAs Soldered with Lead-free Solder Paste. Soldering and Surface Mount Technology, 14 (2): 35-39.
26. Hunt C and Wickham M (2006) Impact of Lead Contamination on Reliability of Lead Free Alloys. In: Proceedings of IPC Printed Circuits Expo, APEX and the Designers Summit, Anaheim, CA, Feb. 8-10, S39-01-1 -11.
27. Practical Components webpage at http://www.practicalcomponents.com/
28. Topline webpage at http://www.topline.tv/
29. Moon K, Boettinger WJ, Kattner UR, Handwerker CA, and Lee D (2001) The Effect of Pb contamination on the Solidification Behavior of Sn-Bi Solders. Journal of Electronic Materials, 30 (1), 45-52.
30. JEDEC/IPC (2006) Current Tin Whiskers Theory and Mitigation Practices Guideline. March.
31. Henshall G, Roubaud P and Chew G (2002) Impact of Component Terminal Finish on the Reliability of Pb-free Solder Joints. In: Proceedings of the SMTA International Conference.
32. Lau J, Hoo N, Horsley R, Smetana J, Shangguan D, Dauksher W, Love D, Menis I, Sullivan B (2004) Reliability Testing and Data Analysis of Lead-free Solder Joints for High-density Packages. Soldering and Surface Mount Technology, 16 (2): 46-68.
33. Seelig K and Suraski D (2001) Advances Issues in Assembly: Part 1 Lead Contamination in Lead-free Assembly. In: SMT Magazine, October.
34. Solectron internal data.

35. Verhelst E and Ocket T (2002) Lead-free manufacturing: Effects on press-fit connections. White Paper, Tyco Electronics.
36. Pal I, Smolentseva E (2006) An Experimental Study of Press-fit Interconnection on Lead Free Plating Finishes. In: Proceedings of SMTAI, Chicago, IL.
37. Chou GJS and Hilty RD (2003) Effects of Lead-free Surface Finishes on Press-fit Connections. In: Proceedings of the IPC Annual Meeting, Minneapolis, MN.

Chapter 8: PCB Laminates

Karl Sauter, Sr. Staff Engineer, Sun Microsystems, Engineering Technologies, Menlo Park, CA

8.1 Introduction

Printed wiring board manufacturing yields and subsequent product reliability are significantly impacted by the properties and characteristics of the laminate materials used. The selection of laminate materials is more critical for achieving higher temperature lead-free solder alloy assembly yield targets and long term product reliability requirements. The SAC305 tin-silver-copper alloy (3.0 percent silver and 0.5 percent copper) recommended for lead-free assembly reflow processing and rework has a melting point about 40°C higher than the melting point of eutectic tin-lead.

How printed wiring boards are designed, and associated printed wiring board manufacturing processing, can also affect the appropriate selection of laminate material that will go through higher temperature lead-free assembly processing and rework. Printed circuit boards for large electronics systems, especially at higher frequencies, have large numbers of signal nets with high density trace routing on the signal layers and high subsystem power demand which requires more copper to minimize DC voltage drop. Construction of these printed circuit boards often requires the use of splitting of plane layers among different supply voltages, sometimes mixing power plane shapes with signal traces on the same layer. This design practice can affect the thermal dissipation characteristics of the printed wiring board. This chapter will address how to select laminate materials for higher temperature lead-free assembly processing, which impacts the quality and reliability of the PCB and associated interconnects.

8.2 Types of Stress in Printed Wiring Boards

Several types of stress can have significant effects upon board yield and subsequent product reliability. Eutectic tin/lead solder melts at 183°C, and printed wiring boards with significant thermal mass can require heating up to 235°C (typically 200 to 220°C) for all plated through hole and surface mount solder joints during assembly processing.

The compositions of tin/silver/copper (SAC) alloy used as the reflow alloy for lead-free assembly processing and rework of solder require printed wiring boards to withstand significantly greater thermal stress. Tin-silver-copper (SAC) alloys with 3.0 to 4.0 percent silver and 0.5 to 0.7 percent copper have a melting points that range between 217 and 220°C. Boards fully loaded with components have significantly greater thermal mass and can require heating up to 260 or even 270°C for forming all associated SAC alloy solder joints. The use of a higher preheat temperature and/or longer pre-heat time than is typical for eutectic tin-lead solder alloy assembly processing can reduce the peak assembly temperature required for SAC alloy reflow, potentially mitigating the damage to heat-sensitive components being soldered onto the board [1].

The variety of stresses on a printed wiring board can interact with one another, and along with electrical potential differences when the assembly is functioning can cause electrical failures. Lead-free SAC solder alloy assembly processing of boards therefore requires a significant increase in preheat and peak temperature, and this produces significantly more thermal stress on the printed wiring board laminate material used. The following stresses can interact with one another, and along with electrical potential differences when the assembly is functioning can cause electrical failures.

8.2.1 Thermal Stress

Thermal stresses that occur during printed circuit board assembly and rework processing can be significantly more severe with lead-free SAC alloy assembly processing. Lead-Free assembly profile of medium-sized board is shown in Figure 8.1.

The combination of longer preheat time and higher temperature assembly reflow and rework processing causes additional strain on the copper in plated through holes due to the greater material expansion of the laminate material, which accelerates metal fatigue within interconnect structures.

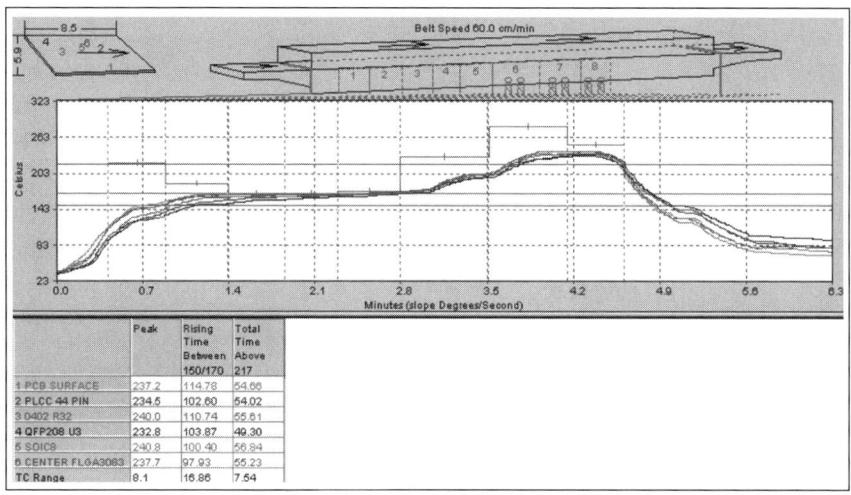

	Peak	Rising Time Between 150/170	Total Time Above 217
1 PCB SURFACE	237.2	114.78	64.66
2 PLCC 44 PIN	234.5	102.60	64.02
3 0402 R32	240.0	110.74	55.61
4 QFP208 U3	232.8	103.87	49.30
5 SOIC9	240.8	100.40	56.84
6 CENTER FLGA3083	237.7	97.93	55.23
TC Range	8.1	16.86	7.54

Fig. 8.1. Medium-Sized Board (6.0 × 8.5 inches, .062" thick) Lead-Free Reflow Profile (Courtesy: Flextronics)

The longer and/or more extreme thermal excursions can cause the dielectric material to delaminate. No one single industry test method or laminate material characteristic adequately predicts both types of thermal stress failures. Accelerated thermal cycling testing of representative samples is performed as an indicator of the combined effect of these factors.

8.2.2 Mechanical Stress

Mechanical stresses on a board can significantly affect assembly yields and long-term product reliability. Standard epoxy-FR4 laminate materials are already somewhat brittle, and fracture within a printed wiring board usually occurs first at or along epoxy fiberglass interfaces. In addition to poor handling practices contributing to fracture, higher temperature lead-free SAC solder alloy assembly and rework processing also can increase mechanical stress failures since more thermally robust laminate materials tend to be harder and more brittle. Mechanical and thermal stresses can interact with one another, and along with electrical potential differences when the assembly is functioning, cause electrical failures.

8.2.3 Chemical Stress

Chemical sources of stress (residual flux, moisture, smog, corrosive gases) in the product service environment are not significantly different with higher temperature lead-free versus eutectic tin-lead solder alloy assembly processing. However during higher temperature lead-free assembly reflow processing the moisture content within the laminate material is much more likely to escape through the plated through holes compared with eutectic tin-lead assembly reflow processing. Out gassing within plated through holes can significantly reduce the reliability of small plated through holes (see Case Study Section of this chapter). Higher temperature assembly reflow processing can also degrade the chemical bonding strength between the copper and epoxy, and reduce the copper peel strength.

8.2.4 Copper Dissolution and Plated Through Hole Reliability

The SAC solder alloys during reflow and wave soldering can dissolve copper much faster than eutectic tin-lead solder alloy, especially at higher temperatures. The SAC305 alloy (96.5Sn3.0Ag0.5Cu) is particularly aggressive towards copper and at 0.5 percent copper content it is well under the saturation limit for copper at 260°C.

Therefore laminate material expansion with temperature, laminate material delamination, bonding strength with copper, and material brittleness are all appropriate material characteristics to consider when selecting a more thermally robust laminate material. Several industry material test methods have some correlation with these characteristics and are used to help determine suitable laminate materials for higher temperature lead-free assembly processing.

8.3 Laminate Material Test Methods

IPC-4101B, "Specification for Base Materials for Rigid and Multi-layer Printed Boards," and its associated material type specifications or "slash sheets," have been the standard for defining the base laminate material properties required by the industry as a whole. For eutectic tin-lead assembly processing only the glass transition temperature of standard FR4 laminate material has been considered critical for the assembly of larger or thicker multi-layer boards with good yields and sufficient product field life. This is not the case for lead-free assembly processing.

8.3.1 Glass Transition Temperature (Tg)

A material's glass transition temperature (Tg) is determined using TMA (IPC TM-650 2.4.24C) and/or DSC (IPC TM-650 2.4.25C) methods of analysis. Tg(DSC) results are typically higher than Tg(TMA) results, so it is important to use the same test method for comparing different materials. Laminate materials with 170°C Tg typically have a lower overall Z-axis expansion during assembly reflow processing than laminate materials with 140°C Tg. The IPC 4101B laminate material slash sheets associate lead-free assembly processing with a laminate material Tg(TMA) of at least 140°C. However thicker boards going through higher temperature lead-free assembly processing are more at risk for delamination and other defects [2,3,4]. Since different laminate materials with varying amounts of filler can have different rates of thermal expansion above the Tg, it is critical to know both the thermal expansion coefficient Z-axis (CTE-Z) expansion rate from ambient to the Tg, and the CTE-Z from the Tg to the actual peak reflow temperatures. Laminate materials that have a higher Tg do tend to be more brittle.

8.3.2 Laminate Material Delamination Testing

Thermal Mechanical Analyzer (TMA) testing measuring minutes to delamination at 260°C ("T-260" testing) has been widely used as a good indicator of the required thermal robustness of laminate materials that will go through eutectic tin-lead assembly processing and rework. Test procedures are specified in IPC TM-650, Section 2.4.24.1, and measure the time in minutes for an irreversible change in thickness at the designated peak temperature. Time to delamination testing at 288°C ("T-288" testing) is more indicative of the thermal robustness required for higher temperature lead-free assembly processing and rework using SAC alloys than the T260 test method. The IPC 4101B laminate material "slash sheets" recommended for lead-free assembly processing do specify minimum T260 values of 30 minutes or more, and T288 values of 5 minutes or more.

8.3.3 Coefficient of Thermal Expansion in the Z-axis (CTE-Z)

Coefficient of Thermal Expansion – Z Axis (CTE-Z) testing per IPC TM-650 (2.4.41) is used to evaluate a laminate material's suitability for the assembly of thick boards. An overall Z-axis expansion of more than 3.5 percent during assembly and/or rework thermal excursions can cause excessive mechanical stress on plated through holes and reduce PTH reliability. Lower coefficient of thermal expansion laminate materials are used to withstand higher lead-free assembly processing temperatures without exceeding overall 3.5 percent Z-axis expansion.

8.3.4 Moisture Resistance

Residual moisture present within printed wiring board laminate material contributes to CAF (Conductive Anodic Filament) failures by escaping or out-gassing through plated through hole walls during high temperature assembly reflow processing. Water Absorption and Moisture Resistance test procedures are specified in IPC TM-650, Section 2.6. IPC 4101B laminate material "slash sheets" recommended for higher temperature lead-free assembly processing specify a minimum moisture resistance of 0.5 percent.

8.3.5 Peel Strength

Higher temperature assembly reflow processing can degrade the chemical bonding strength between copper and epoxy, reducing the copper peel strength which can consequently lower rework yields. Peel strength test procedures are specified in IPC TM-650, Section 2.4.8. IPC 4101B laminate material "slash sheets" recommended for lead-free assembly processing specify minimum peel strength values of 0.70 N/mm (4.0 lb/inch) or more at 125°C.

The IPC-4101-B has additional slash sheets and additional material property testing requirements useful for evaluating laminate materials for boards that will through higher temperature lead-free assembly processing and rework. Key parameters potentially critical for selecting more thermally resistant laminate material include, in addition to Tg, T260, and CTE-Z (percent change over the full temperature range from ambient to peak reflow temperature); Td (Laminate Material Temperature of Decomposition) and minutes to delamination during T288 testing. T288 results are more applicable to higher temperature lead-free SAC alloy assembly processing than T260 results. The additional laminate material test methods incorporated in the 4101-B "slash sheets" recommended for lead-free assembly processing include the Td test method.

8.3.6 Laminate Material Temperature of Decomposition (Td) Test Method

The Laminate Material Temperature of Decomposition (Td) Test Method (IPC TM-650, Section 2.4.24.6) is a measure of the thermal robustness and/or interlamination bond strength of a laminate material. The development of this test method (TGA, 5% weight loss) involved the cooperation of several major laminate material suppliers and OEMs in a round robin study that concluded in the gage R&R testing of a large number of samples. The gage R&R results contributed to the acceptance of the Td test method in the IPC-4101B specification. The more thermally robust phenolic-cured laminate materials more suitable for higher temperature lead-free assembly processing generally have a laminate material decomposition temperature of at least 325°C.

8.3.7 Conductive Anodic Filament (CAF) Resistance Test Method

The Conductive Anodic Filament (CAF) Resistance Test Method (IPC TM-650, Section 2.6.25) and the associated CAF Resistance Test Method User Guide (IPC-9691) are useful for characterizing more thermally robust laminate materials used in higher temperature lead-free assembly processing. This test method is sensitive to the extent of laminate material fracturing around drilled and plated through holes [5]. The IPC CAF Test Method Users Guide recommends a CAF testing pass/fail criteria for laminate material used in higher temperature lead-free assembly processing. An example of a CAF Failure is shown in Figure 8.2. Note: Checking for "dry weave" or "interfilament separation" within the laminate material is recommended to identify poorly processed laminate material prior to building representative test boards for CAF testing.

CAF testing is important for evaluating materials that will go through higher temperature lead-free assembly processing because the associated more severe thermal excursions will propagate further the micro cracks normally present around the holes drilled in multi-layer boards. Printed wiring board processing chemicals and other ionic materials tend to collect in these fractured areas, and contribute to several types of electrochemical migration resulting in reduced isolation resistance failures.

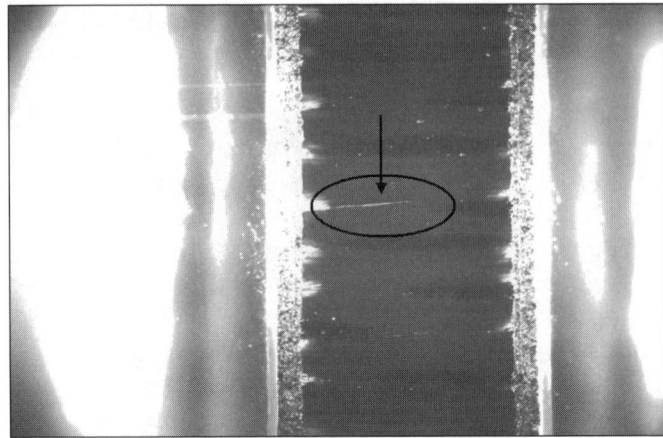

Fig. 8.2. CAF Failure (formation along resin/fiberglass interface) – IPC CAF Test Board

8.4 Accelerated Thermal Stress Testing

Laminate material characteristics vary, so selecting suitable laminate materials for higher temperature lead-free assembly processing is a more complex process. There are a variety of higher temperature lead-free alloy reflow profiles, rework processes, and types of board assembly. For a limited set of printed circuit board characteristics it may be possible to develop a higher temperature lead-free assembly performance index or predictive model. This index would be a function of assembly parameters and laminate material characteristics that correlate with assembly yields and long term product reliability (i.e. peak assembly reflow temperature, preheat, Tg, Td, CTE-Z, board thickness, CAF). Engelmaier has proposed one such an index [6]. However acceptance of any index for determining laminate material suitability for a given higher temperature lead-free assembly and rework process should require comprehensive long term thermal cycling test data and/or actual product field data as well as assembly yield data.

8.4.1 Critical Factors in Accelerated Thermal Stress Testing

The selection of laminate materials for higher temperature lead-free assembly processing requires accelerated thermal stress testing. Meaningful test results require that all contributing factors be considered when developing an accelerated thermal stress test plan as a good indicator of long term product reliability. The factors presented are critical for cost-effective and accurate correlation of accelerated thermal stress test results with long-term product reliability.

8.4.1.1 Initial Sample Selection

The first samples selected for conducting thermal cycle testing should be from or representative of a well understood volume production product made with the standard volume production laminate material for which there is ample field data regarding long term product reliability. Standard PWB design features and multi-layer printed wiring board manufacturing should be sufficient for surviving eutectic tin-lead assembly processing. Other characteristics may be required for representative first test boards to serve as a reliability benchmark or standard. These samples for thermal cycle testing can also establish a baseline when first using a new thermal cycling test method.

8.4.1.2 Test Board Preconditioning

Representative test board preconditioning is required prior to thermal cycle testing since assemblies in production can go through different profiles and/or allow varying degrees of rework depending upon the on-board component types, overall board thickness and attachment method. Representative preconditioning consists of thermal excursions conducted to match the actual board assembly peak reflow temperatures used and the anticipated rework processing. Two preconditioning thermal excursion cycles are added for each anticipated component rework cycle, one for component removal and the second for attaching a new component at the same location. Samples that are preconditioned go through the expected or worst case assembly thermal excursions that the actual product will go through. Lead-free assembly processing temperatures (preheat time and temperature, peak temperature) vary significantly depending upon the component mix, component placement density, overall board thickness, size, shape, thermal mass distribution, and other factors.

8.4.1.3 Alternate Laminate Materials

Alternate laminate materials should be tested only after testing is completed on a standard material since cost-effective testing to estimate new material impact on long term product reliability is based upon a relative comparison with test boards made using the standard/baseline laminate material. More thermally robust laminate materials can take a very long time to fail under conditions representative of eutectic tin-lead assembly processing. Budgetary constraints limit correlation of long term product field life only with standard material test board accelerated thermal cycle (ATC) test results. Therefore the results of ATC testing of more thermally robust laminate materials combined with higher temperature lead-free assembly profiles are cost-effectively evaluated by comparing those results with standard baseline test board results.

8.4.2 Sample Design

Representative samples that go through ATC testing after appropriate thermal excursion preconditioning should have design features as similar as possible to the actual product. Layer count, board construction and overall board thickness are primary considerations. A printed wiring board change from lower to higher aspect ratio plated through holes (3.0 mm overall thickness, ~0.3 mm DHS (drilled hole size) and with higher via density (high I/O 1.0 mm BGA devices) can significantly impact accelerated thermal cycle test results. The use of more thermally robust laminate materials introduces another significant variable. Drilling, desmear and other PWB manufacturing processes should be adjusted for a new laminate material in order to achieve the PTH quality necessary for optimum accelerated thermal cycle test results.

8.4.3 Thermal Cycling Limits

Subsequent long term thermal cycling (reliability testing) should be conducted under conditions that accelerate the dominant failure mechanism without introducing new failure mechanisms or failure mechanism interactions with other material characteristics that are not significant under actual field conditions. For example, FR4 laminate materials require thermal cycle testing with a maximum temperature below the laminate material Tg. This assures that the long term thermal cycling test results are more a function of plated through hole quality, percent copper elongation property and the geometry of the plated through hole. At higher cycling temperatures near and above the laminate material Tg, the modulus of the laminate material can change significantly resulting in thermal cycling failure rates more influenced by the type of laminate material formulation, PTH geometry, board design and processing variables.

8.4.4 Field Life Correlation

Due to the number of significant variables involved, accelerated thermal cycling test results for FR4 laminate materials are primarily used to qualify or monitor product as an indicator of the relative product reliability. Long term thermal cycle testing can also be done to develop a correlation with actual long term product reliability in the field. However, to be representative, this requires testing the same product at several different peak cycling temperatures, which can be very expensive.

8.5 Accelerated Thermal Stress Test Methods

8.5.1 Oven Testing

Oven testing is the oldest of the accelerated thermal stress test methods used by the printed circuit board industry, and produce a very uniform cyclic strain within the interconnect, potentially resulting in more representative acceleration of the particular failure mechanism(s) that cause product to fail in the field. Thermal excursions within the oven or test chamber should produce uniform strain within the material tested. Printed wiring board structures such as plated through hole interconnects distribute and otherwise react to the thermally induced mechanical strain, and time to failure is regarded as an indicator of product reliability in the field.

Plated through hole barrels and junctions with internal traces and plane layers are repeatedly stressed until failure during ATC testing. Classic metal fatigue is generally the failure mechanism associated with high quality printed wiring boards, and these failures are typically located near the center of the plated through hole or near the 'knee' of the plated through hole (the site of repeated surface pad lifting). However oven thermal cycling test chambers can be difficult to manage in terms of ensuring good airflow and uniform humidity without condensation during ATC testing. In addition the cycle times are longer, sometimes requiring several months of testing before a fifty percent failure rate is reached.

8.5.2 IST Testing

Interconnect Stress Testing (IST) is the most widely used method for monitoring plated through hole (PTH) reliability. IST allows local current heating for representative preconditioning and subsequent thermal cycling where IST coupons are typically cycled from ambient to 150°C for a minimum number of cycles or until test coupons reach a fifty percent failure rate. Cycling below the laminate material Tg provides a consistent measure of the PTH integrity, and the presence and level of post interconnect type separation failures in multi-layer boards can be identified. Due to shorter test cycle times the frequent use of IST compliments and can possibly reduce the levels of cross-section analysis required for monitoring PTH quality/reliability.

IST results show that both plated through hole (PTH) and post interconnect failures can precipitate as independent failure mechanisms, or can interact to accelerate interconnect failures. Along with HATS testing (reference next section), IST is a proven accelerated reliability test method and can be used to monitor different levels of interconnect degradation as they occur.

8.5.3 HATS Testing

HATS (highly accelerated thermal stress) test can be a practical choice for determining long-term reliability with an example of a HATS test chamber shown in Figure 8.3.

Fig. 8.3. HATS Test Chamber (courtesy Microtek Labs)

HATS testing offers a wide thermal cycling temperature range below the Tg which avoids introducing new PTH failure mechanisms or significantly changing the actual product field life failure mechanism. The IST is limited to ambient (+25°C) at the low end, while HATS thermal cycling can go as low as –55°C.

HATS test setup includes the capability of a controlled and not too rapid temperature rate of rise, and rate of cooling (air-to-air heating and cooling). Studies have shown good correlation between HATS test results and Thermal Shock oven test results for standard dicyandiamide-cured FR4 laminate material with peak cycle temperatures below Tg.

The Delphi Reliability Study[7] showed good agreement between all three thermal cycling test methods as indicated in Figure 8.4. For standard dicyandiamide cured 170°C Tg FR4 laminate materials with a peak cycling temperature of 145°C (not too close to the Tg), a 50 percent failure rate corresponds to 180 +/– 20 cycles to fail. For more thermally robust phenolic cured 180°C Tg FR4 laminate materials and the same temperature cycling, a 50 percent failure rate corresponds to between 630 to 770 cycles to fail (700 cycles +/– 10 percent).

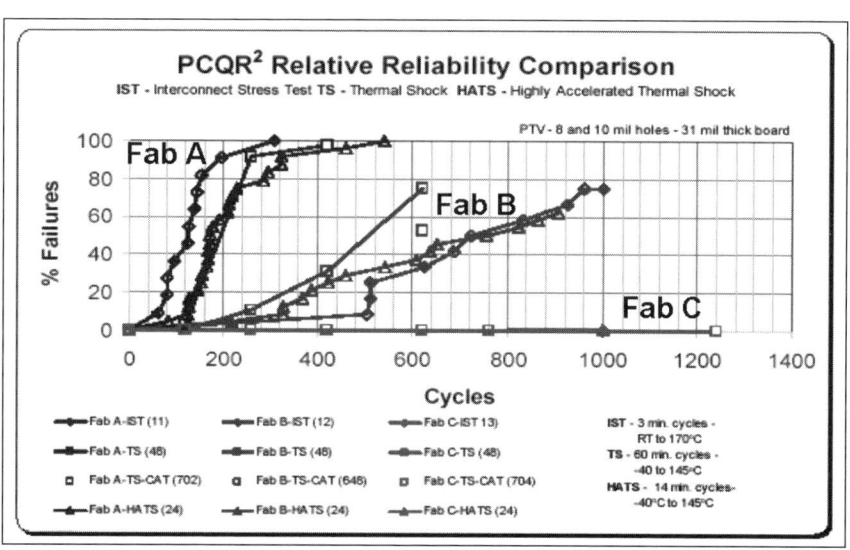

Fig. 8.4. IPC PCQR2/CAT - Delphi Reliability Study (David L. Wolf, 14 April 2003) [7]

8.6 HATS Test Method – A Case Study

This test method was used to evaluate the effects of different ATC thermal cycles and levels of preconditioning. The test plan parameters were:

Preconditioning peak reflow temperature = 230°C
36 coupons per chamber load, coupon size 1.0 × 2.0 inches
4 daisy chain nets per coupon (2 drill sizes, 2 daisy chain routing structures).182 plated through hole vias per daisy chain net
Number of preconditioning cycles = 0X, 4X, and 8X

Thermal Cycle Type #1 = 25°C to 160°C (for comparison with IST test results)
Thermal Cycle Type #2 = 0°C to 135°C (same temperature range as Type #1)
Thermal Cycle Type #3 = 0°C to 160°C (larger delta temp to shorten test time)

The average values of the Case Study HATS test results had greater variability than was expected based upon earlier work, as shown in Table 8.1.

Table 8.1. HATS Via PTH Reliability Testing Case Study and Results

Average Cycles to 10%Increase in Resistance	Layers 1-26, Drill 13.5 (1)	Layers 1-26, Drill 11.5 (2)	Layers 4-23, Drill 13.5 (3)	Layers 4-23, Drill 11.5 (4)
(0-160) 8X	579	440	406	365
(0-160) 4X	760	548	506	472
(0-160) 0X	730	506	526	486
(25-160) 8X	720	600	551	503
(25-160) 4X	860	724	592	597
(25-160) 0X	1,050	790	830	730
(0-135) 8X	>3,000	>3,000	>3,000	>3,000
(0-135) 4X	>3,000	>3,000	>3,000	>3,000

Data Set #1: Daisy chains routed from layer 1 to 26, and drilled with 13.5 mil drill bit diameter.
Data Set #2: Daisy chains routed from layer 1 to 26 and drilled with 11.5 mil drill bit diameter.

Data Set #3: Daisy chains routed from layer 4 to 23 and drilled with 13.5 mil drill bit diameter.
Data Set #4: Daisy chains routed from layer 4 to 23 and drilled with 11.5 mil drill bit diameter.

None of the four HATS test net data sets had a relationship between the number of preconditioning cycles and the HATS testing cycles to failure (CTF), nor correlation with either Miner's Rule or the Inverse Power Law. The HATS test results when cycling from 0 to 135°C took a much longer time to fail even though the HATS delta T of 135°C was the same as for the previous set, indicating a complex failure mechanism. This data was not sufficient to demonstrate a relationship between number of preconditioning cycles and cycles to failure.

The analysis of these test results are shown below in Figure 8.5, which compares the target minimum long-term product reliability (arrowhead pointing to the right) with the reported HATS test results.

Although the nominal reliability appears sufficient, the great variability in the two delta 135°C test results (estimated stress 4200 – 4500 ppm) did not allow an accurate correlation with long term field life. Therefore root cause failure analysis was conducted to help determine the most probable cause(s) of this variability.

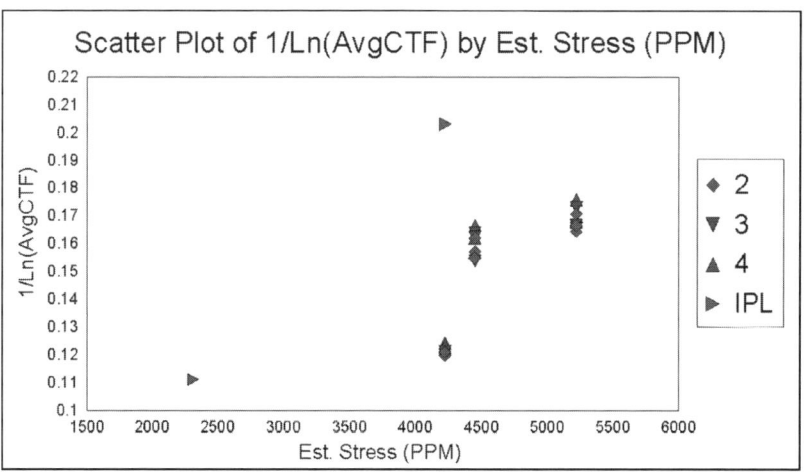

Fig. 8.5. Minimum 25 Year Reliability Target (IPL) and HATS Test Results (Panel #)

8.6.1 Root Cause Analysis of Failed Nets

The ideal plated through hole (PTH) failures during long term reliability thermal cycling testing are classic metal fatigue failures and are characterized by circumferential cracking around the PTH located approximately equidistant between the top and bottom layers. The photograph shown below in Figure 8.6 shows how the copper remains fairly close together after cracking.

Based upon the findings of this case study, the cross-sections of the failed HATS test coupons were examined for evidence of the copper metal fatigue failure mechanism. Cross-sections were taken from every HATS test group, and all test groups showed failures that were not characteristic of classic metal fatigue failure, as shown in Figure 8.7.

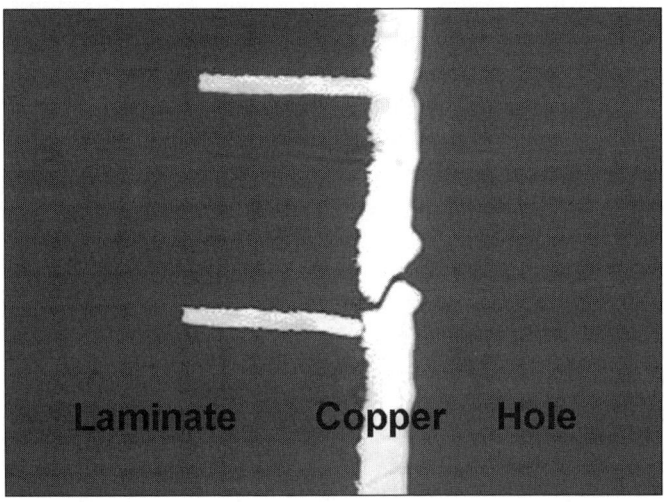

Fig. 8.6. Ideal Thermal Cycling Failure is a Classic Metal Fatigue Failure within the PTH. Courtesy: Sanmina-SCI, Phoenix, AZ

The actual cross-sections at the failure sites showed evidence of out gassing at various points along both sides of the plated through hole wall and at several locations, not just in the middle of the hole, and possibly associated delamination within the laminate material. Although out gassing prior to HATS testing can cause the early initiation of PTH cracking, out gassing sites inside a plated through hole can also stress relieve the remaining copper along the hole wall resulting in much longer cycles to failure if subsequently cycled at only lower temperatures.

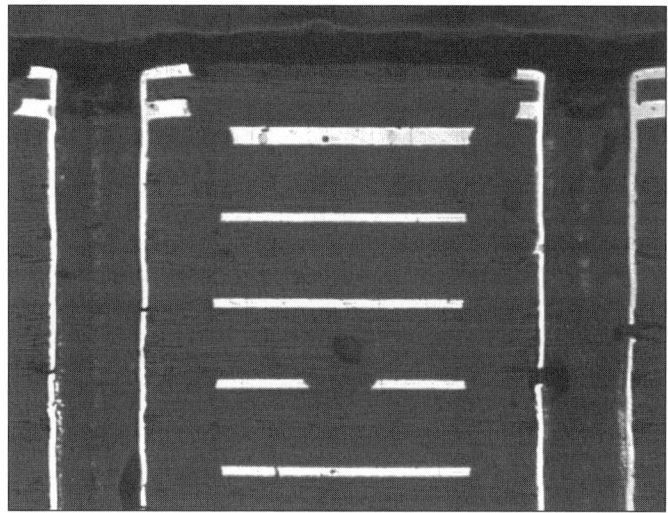

Fig. 8.7. Typical cross-section of the Case Study Test Failures

In conclusion, since the plated through hole (PTH) out gassing was observed on all samples not just the 0 – 135°C samples, the out-gassing alone is not the reason for the difference in cycles to failure between the 0 – 135°C and 25 – 160°C test groups. The HATS testing temperature ramp rate for the 0 – 135°C test group was significantly more rapid than for the 25 – 160°C test group, and in combination with the observed out gassing sites this could explain the significant difference in cycles to failure between the two 135°C delta T test groups. The presence of PTH out-gassing is therefore associated with earlier times to failure when higher peak temperatures for thermal cycling are used.

Figure 8.8 compares the results of several published thermal cycling test methods, which have been used to estimate long term product reliability with the above case study. The double triangle shows the results of a HDPUG study [8]. Note that the HDPUG test results just barely exceeded standard Telecom long term board reliability requirements (indicated by a straight diagonal line between the two triangles that point to the right).

The results from the Delphi study [7] are indicated by the two triangles that point to the left. Although the Delphi results seem to be in agreement with these HATS Case Study results, the Delphi samples were not precon-ditioned and there was no checking for the presence of out-gassing within the plated through hole. The HDPUG study [8] samples were made with the

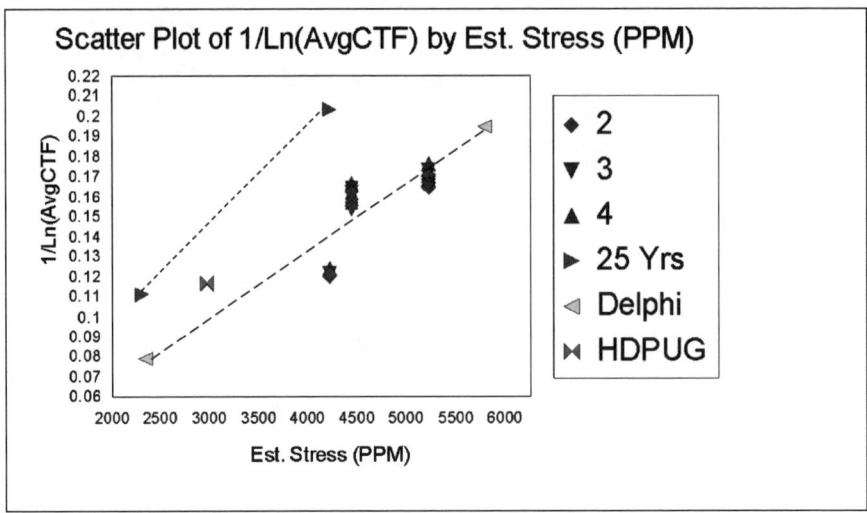

Fig. 8.8. Minimum 25 Year Reliability Target (IPL) and HATS Test Results (Panel #)

same high Tg dicyandiamide cured FR4 laminate material and also went through preconditioning representative of eutectic tin-lead assembly and rework processing.

The case study described herein and the referenced published reports emphasize the importance of well-conducted and representative thermal cycling testing to determine the impact of using alternative laminate materials on long term product reliability. In general the published data for printed wiring boards made with dicyandiamide cured FR4 laminate material do not show a comfortable margin for error for projecting reliability over 20 years of product life in the field.

The following factors should be considered in order to ensure meaningful thermal cycle testing and results for determining long term product reliability:

i) Changes in board design and in printed wiring board manufacturing processes can affect the appropriate selection of laminate materials. Therefore both representative test samples and representative preconditioning (simulating thermal excursions that do or will occur during actual board assembly and rework) should be used when conducting thermal cycle testing.

ii) Include some test samples preconditioned to match as closely as possible the actual assembly processing and rework thermal excursions. For example, the actual lead-free peak reflow temperature used can vary based upon solder paste used, reflow alloy used, board size, thickness, and other characteristics.

iii) Ensure samples are sufficiently dry both prior to preconditioning and on-going thermal cycling testing that is done below the laminate material Tg.

iv) Always cross-section failed samples for root cause failure analysis to determine the failure mode. The desired failure mode is fatigue failure (characteristic of long term field life reliability modeling).

v) Destructively test and check a few samples prior to starting the on-going thermal cycling testing to ensure that no defects are present that would drive other failure modes (examples: dry weave, delamination, excessive moisture).

vi) Ensure temperature rate of rise and rate of cooling are controlled in the same way (same rate of temperature change) for all sample groups to prevent this being a cause of excessive variability in the test results.

Cost is a factor when selecting materials, and several types of more thermally robust laminate materials should be evaluated if more than one material is able to withstand the given board assembly process and meets the end product reliability requirements. A lower cost laminate material can then be selected from the materials able to withstand the given board assembly process and meet end product reliability requirements.

Lead-free assembly and rework peak temperatures do vary based upon board requirements, and can range from 230°C to 270°C. Assembly processing that requires over 245°C peak reflow temperature and/or higher and longer preheat temperatures may require the use of a more thermally robust laminate material. Phenolic and aromatic cured laminate materials generally are more thermally robust than dicyandiamide cured FR-4 laminate materials. Boards 2.5 mm (0.100 inches) thick or more and/or which go through assembly or rework processing temperatures approaching 260°C or more may require laminate material containing higher levels of filler or even engineered filler in order to control Z-axis thermal expansion

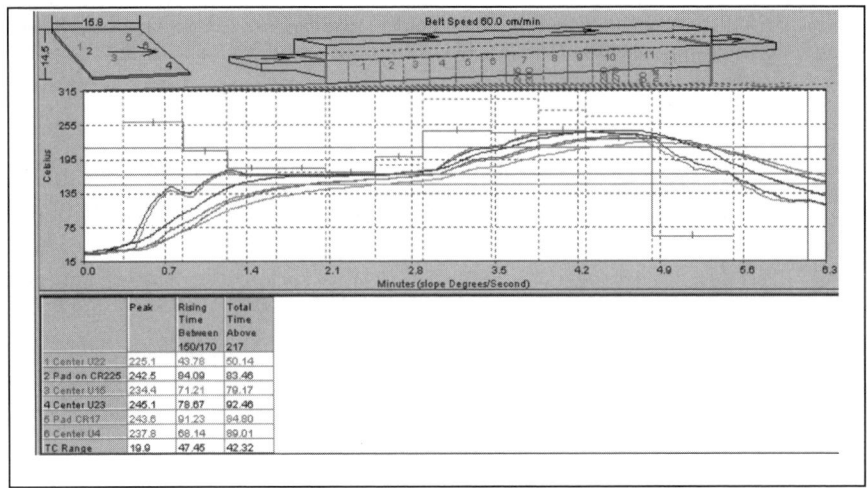

	Peak	Rising Time Between 150/170	Total Time Above 217
1 Center U22	225.1	43.78	50.14
2 Pad on CR225	242.5	84.09	83.46
3 Center U16	234.4	71.21	79.17
4 Center U23	245.1	78.67	92.46
5 Pad CR17	243.6	91.23	84.80
6 Center U4	237.8	68.14	89.01
TC Range	19.9	47.45	42.32

Fig. 8.9. Large Board (14.0 × 16.0 inches, .092" thick) Lead-Free Reflow Profile (Courtesy: Flextronics)

and meet end product reliability requirements. An example of this higher temperature lead-free assembly reflow profile is shown in Figure 8.9.

Although several laminate material test methods are useful for characterizing a laminate material's thermal robustness, no single material characteristic alone is a good indicator for successful higher temperature lead-free assembly processing and rework. One material set and assembly peak reflow process will not be suitable for all market segments, or may need to be changed if a product falls outside an established mainstream application. Laminate material characteristics vary in significance depending upon the actual peak assembly temperature used, and the number and type of subsequent rework cycles. For example the use of laminate material with engineered fillers (more controlled sizing, purity, and mix) may be required only for larger, thicker boards going through higher temperature lead-free assembly and/or rework processing.

Conclusions

There are several laminate material properties potentially critical for evaluating laminate material choices for higher temperature lead-free assembly processing. The selection of laminate materials for higher temperature lead-free assembly processing therefore requires specifically representative accelerated thermal stress testing due to the variety of factors potentially impacting end product quality and reliability. Laminate material decomposition temperature is the newest test method for thermal robustness and along with overall Z-axis CTE these two methods are the most indicative of what is required for both surviving higher temperature lead-free assembly processing and ensuring subsequent product reliability. Time to delamination test results are more indicative of assembly survivability than of subsequent product reliability.

Future Work

Low pressure areas present during the press lamination cycle of printed wiring board manufacturing are more prone to subsequent delamination during higher temperature lead-free assembly processing. Several OEMs and industry groups are working on the development of test methods that can better predict the risk or level of occurrence of these defects, particularly for larger and thicker multi-layer boards going through longer preheat times prior to actual peak lead-free assembly reflow processing.

Acknowledgements

The author would like to acknowledge the following persons for suggestions and information in writing this chapter:

David Love, Sun Microsystems
Jasbir Bath, Solectron
Joe Smetana, Alcatel
Werner Engelmaier, Engelmaier Associates
David Geiger, Flextronics

References

1. Baggio T, Suetsugu K (1999) Guidelines for Lead-Free Processing, SMT magazine
2. Davignon J, Reed R (2000) Effects of NEMI Sn/Ag/Cu Alloy Assembly Reflow on Plated Through Hole Performance, NEPCON East Conference, Boston, USA
3. Smetana J (2002) Plated Through Hole Reliability with High Temperature Lead-Free Soldering, The Board Authority magazine

4. Leys D, Schaefer S (2003) PWB Dielectric Substrates for Lead-Free Electronics Manufacturing, CircuiTree magazine
5. Turbini L, Bent W, Ready W, Impact of Higher Melting Lead-Free Solders on the Reliability of Printed Wiring Assemblies, Center for Microelectronics Assembly and Packaging, University of Toronto, Toronto, Canada, and Georgia Institute of Technology, Atlanta, USA
6. Engelmaier W (2003) Reliability of Lead-Free Solder Joints Revisited, Global SMT & Packaging, p. 34
7. Wolf D (2003) Delphi Reliability Study, IPC PCQR2/CAT
8. Smetana J (2006) Via (PTH) Integrity with Lead Free Soldering, IPC PC EXPO-APEX conference, Anaheim, USA, S24; http://www.hdpug.org/public/4-papers/apex-2006/via-integrity-paper.pdf

Chapter 9: Lead-Free Board Surface Finishes

Hugh Roberts, Atotech USA Inc., Rock Hill, SC, USA
Kuldip Johal, Atotech USA Inc., Rock Hill, SC, USA

9.1 Introduction

In the move to lead-free electronics, one of the main aspects of printed wiring board (PWB) fabrication that has received attention is the surface finish. The surface finish of the PWB represents the last major step in fabrication before component assembly and, as such, represents the interface between the external board circuitry and the bonding medium (i.e. solder).

The use of lead-free solders requires higher assembly temperatures and places increased demands on the surface finish if it is to survive multiple reflow cycles. In PWB fabrication, the surface finish can serve several inter-related functions, including:

- Providing a solderable coating to form a strong solder joint
- Protecting the underlying copper circuitry from oxidation or other forms of chemical corrosion until downstream assembly operations are performed
- Offering a medium that is suitable for certain bonding operations during assembly of components (i.e. wire bonding or press-fit)
- Providing a barrier to minimize copper dissolution during assembly operations
- Serving as a functional interface for final product use (i.e. contact switching operations)

Historically, the surface finishes of the PWB incorporated lead (Pb) in their composition, whether in the form of electroplated tin/lead, or more commonly as hot air leveled solder (HASL). The surface was typically comprised of the eutectic (63/37) Sn/Pb matrix. With the increasing acceptance of surface mount technology in the early 1990s, the technical limitations of HASL were being recognized and were overcome with the introduction of new processes that provided better co-planarity. Deposits that were created by electroless or immersion technologies provided a flatter surface and conformed more exactly to the topography of the underlying plated copper, thus meeting the SMT requirements. However, for many mainstream applications, HASL continued to provide the most reliable and cost-effective assembly solution. Now, for many electronic products, meeting the requirements of the Pb-free initiatives that have recently taken effect in many parts of the world (e.g. RoHS) necessitates the increased use of alternatives to eutectic tin-lead HASL finishes.

In this chapter, various surface finish methodologies are presented and examined. An attempt has been made to establish some guidelines regarding finish specifications, such as deposit thicknesses or composition. In most cases, such specifications are normally established by the end users (i.e. ODM/OEM) and a wide variation in requirements can and does exist. There is also an attempt to establish a range of parameters for the various surface finish fabrication processes, such as sequence steps, treatment times and treatment temperatures. Again, it is important to recognize that, depending on process supplier, some variation exists and that these parameters are provided as examples only.

9.2 Process Overview

A number of surface finishes are being implemented or considered by PWB fabricators to meet the requirements of the Pb-free initiatives, including:

- Hot Air Solder Leveling (HASL)
- Organic Solderability Preservative (OSP)
- Immersion Silver (Imm Ag)
- Immersion Tin (Imm Sn)
- Electroless Nickel/Immersion Gold (ENIG)
- Electroless Nickel/Electroless Palladium/Immersion Gold (Ni/Pd/Au)
- Electrolytic Nickel/Electrolytic Gold
- Electroless Nickel/Immersion Gold/Electroless Gold (ENIGEG)
- Direct Immersion Gold (DIG)
- Custom Finishes

The following is a brief discussion of each of these processes, including deposit information, treatment steps and key issues.

9.2.1 Hot Air Solder Leveling

For many years, the hot air solder leveling (HASL) process served as the process of choice for many PWB assembly applications. The surface finish essentially consisted of the same alloy used for the soldering of components and therefore provided the best metallurgical match for both wave soldering and solder paste reflow techniques. However, with the move to Pb-free assembly, the Pb-free alloy that must be used in the HASL process is also quite different. The use of Pb-free HASL requires some key modifications in the process, such as [1]:

- Higher process temperatures for lead-free solder
- Longer contact time
- Improved alloy circulation for better heat transfer
- Pre-heating the panel (pre-dip)
- High temperature resistant chemistries (oils and fluxes)
- Increased alloy composition control (due to higher dissolution of copper in Pb-free solder)

With alloys that were not really suited for the process, results of early Pb-free HASL efforts were not overly promising. Often, the high temperatures required to achieve proper fluidity of the alloy would cause damage to the PWB. Pb-free HASL is reported to operate within a temperature range of 265-280°C. However, a nickel-modified tin-copper eutectic alloy was found to work well in the HASL process at temperatures of 260-265°C, less than that required for the unmodified eutectic Sn-Cu solder. According to published reports, adding nickel in concentrations of approximately 0.06% by weight achieves a Sn-Cu-Ni HASL coating that is superior in terms of the thickness uniformity to that typically obtained with eutectic Sn/Pb solder [2,3]. Figure 9.1 presents some examples of the surface finish achieved with the Sn-Cu-Ni alloy.

Fig. 9.1. Examples of Pb-free HASL surface finish with Sn-Cu-Ni alloy (Source: Nihon Superior Co. Ltd)

Deposit Characteristics

Lead-free HASL can be used for those applications currently being met by eutectic HASL and the uniformity of the coating thickness is reportedly improved [4]. Similar to the eutectic tin-lead version, use of Pb-free HASL poses uniformity difficulties for SMT and fine-geometry applications. For example, as shown in Figure 9.2, the typical working range of the thickness is 1.75-15.0 μm (70-600 μin) across the board. Also, through-hole diameters are usually reduced by at least 25 μm (1.0 mil). However, such variance in surface planarity is still not suitable for many SMT applications. In addition, the quantity of solder deposited on I/Os and PWB features can be excessive for the solder joints of fine-pitch packages, resulting in shorts between leads and increasing the difficulty of inspecting the solder joints [5]. The following table summarizes results of recent testing of Pb-free alloy thickness across a range of features including large ground areas using a common test panel [1]

The deposit thickness of Pb-free HASL may vary, depending on the specific alloy used. Table 9.1 presents information regarding the range in deposit thickness for some common Pb-free HASL alloys.

Table 9.1 Thickness Range of Common Pb-free HASL Deposits

Pb-free Solder Alloy	Thickness Range (μm)
Sn-0.3%Ag-0.7%Cu	2.6 - 14.2
Sn-3%Ag-0.5%Cu	1.0 - 12.3
Sn-0.7%Cu-0.06%Ni	2.7 - 14.7

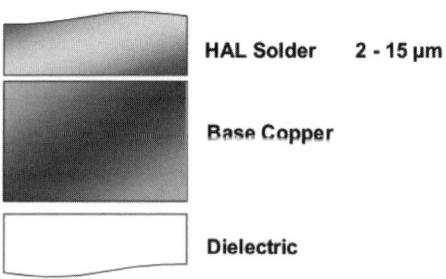

HAL Solder   2 - 15 μm

Base Copper

Dielectric

Fig. 9.2. HASL deposit showing relative layer thicknesses

Process Description

HASL involves applying a solderable alloy finish to the traces and pads of a PWB by first applying flux to the board, preheating it and then immersing it for a few seconds in molten solder. The board is then withdrawn and air knives remove excess solder, leaving a smooth, bright finish [6]. The process is illustrated in Figure 9.3.

For Pb-free applications, the temperature of the molten solder in a HASL system is approximately 260°C. Figure 9.4 shows an example of a Pb-free HASL system.

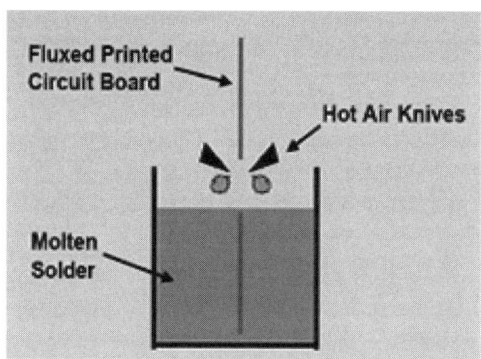

Fig. 9.3. Illustration of HASL operation (Source: Florida CirTech)

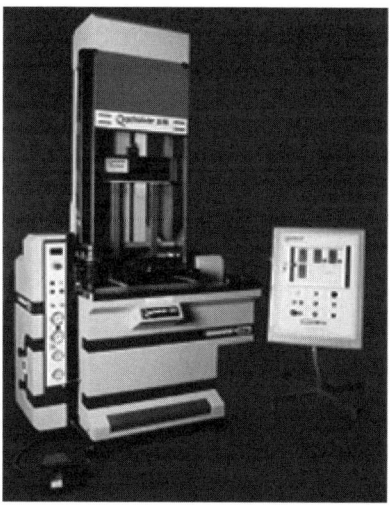

Fig. 9.4. Lead-free vertical HASL machine (Source: Cemco-FSL)

A typical process sequence for Pb-free HASL is presented in Table 9.2.

Table 9.2. Hot Air Solder Leveling (HASL) – Typical Process Sequence (horizontal mode)

Process Step	Treatment Temperature (°C)	Treatment Time	Function
Acid Clean	40-50	30-60 sec	Prepares copper surface for uniform treatment in subsequent microetch step
Rinse	RT		
Microetch	20-30	30-60 sec	Provides defined roughness by exposing the copper grain structure
Rinse	RT		
Flux	n/a	50-70 sec	Applies flux to the PWB prior to immersion in solder
Hot Air Level	260	2-8 sec	Selectively deposits molten solder on the exposed copper surface. Excess removed by air knives.
Post-clean	40-50	30-60 sec	
Rinse	RT		
Dry			

Key Issues - HASL

For many years, HASL has been the principal method for applying a solderable finish to PWBs. The primary reasons for HASL's popularity have been its relatively low cost and its near-perfect compatibility with reflow and wave solder operations (due to their identical compositions). Like its eutectic Sn/Pb version, another advantage of the Pb-free HASL process is its ability to provide an early indication regarding the solderability of the base material surface. Therefore, any soldering deficiencies should be detected before assembly, avoiding more costly rework and repair [5]. However, similar to eutectic Sn/Pb solder, the Pb-free form of HASL poses technological obstacles, such as poor planarity and tendencies to bridge fine-geometry features. As previously mentioned, the deposit thickness on a standard panel can typically vary by as much as 10μm (0.4 mil), creating severe challenges for surface mount applications. There are also processing difficulties pertaining to thin and thick PWBs, both related to the inability of the hot air flow to remove the correct amount of solder from the via hole (i.e. too little solder remaining in very thin PWBs and too much in thick PWBs). Although the use of Sn-Cu-Ni reportedly improves the fluidity and thickness uniformity of the deposit, demanding applications with fine features remain a challenge.

Like eutectic Sn/Pb HASL, the Pb-free HASL process represents additional thermal excursions that the PWB must experience. As a minimum, the temperature during the HASL process will be at least 260°C. In particular, the process also places considerable thermo-mechanical stress on the solder mask. Ultimately, the impact of this thermal excursion is dependent on the PWB design and subsequent assembly requirements.

The higher tin containing lead-free Sn-Ag-Cu alloys for HASL dissolve copper from the board surface at a higher rate than eutectic Sn/Pb HASL. As such, they reduce the copper thickness from pads and holes, most evident at the "knee" of the through-hole. The copper dissolution also makes it very difficult to keep the solder bath composition within specification, where the copper is an alloying element and not an impurity (opposite to the case in eutectic Sn/Pb HASL systems) [6].

Table 9.3 provides an overview of the primary benefits and concerns regarding the Pb-free HASL finish.

Table 9.3. Hot Air Solder Leveling (HASL) – Process Summary

Benefits	Concerns/Limitations
+ Very long shelf life (>12 months)	– Planarity limitations with SMT applications
+ Complete wetting of exposed copper surface	– Limitations with fine feature geometries
+ Suitable for Pb-free soldering	– Compatibility of existing HASL equipment with Pb-free process
+ Well known process	– Creates an additional thermal stress on the board
+ Provides a pre-assembly indication of the solderability of the base material surface	– Dissolution of copper during HASL and subsequent wave soldering
+ Suitable for compliant pin (press-fit) applications (with thickness limitation)	– Not wire-bondable
+ Relatively low-cost	– Not suitable for wear-resistance (contact switching)
+ Ease of optical inspection and ICT	– Limitations with thin or thick PWBs

9.2.2 Organic Solderability Preservative (OSP)

Organic Solderability Preservatives (OSPs) are typically derivatives of benzotriazole, imidazole or benzimidazole that selectively bond to the exposed copper surface to provide protection against oxidation or tarnish. As such, OSPs maintain the solderability of the underlying copper until soldering can occur. In the early 1990s, use of OSPs increased dramatically with the steady growth of surface mount technology. Early versions of OSP were limited in terms of the number of thermal excursions they could withstand, often only a single reflow. However improvements to the products, coupled with the need for OSPs that can withstand the demands of Pb-free assembly, have resulted in development of high-temperature OSPs that exhibit significantly better performance. Presently, OSPs are used for a wide range of applications, including consumer, telecommunications and automotive electronics.

Deposit Characteristics

OSP is primarily composed of organometallic polymer with small molecules, such as fatty acids and azole derivatives, entrapped in the coating during deposition [7]. OSPs are unique in that their properties allow them to bond only with copper. The OSP deposit thickness will typically be in the range of 0.2-0.5 μm (8-20 μin), as shown in Figure 9.5. Multiple reflows in either air or nitrogen atmosphere will reduce its thickness, although reflow in air will reduce the OSP thickness more than reflow in nitrogen atmosphere. The decomposition temperatures for OSPs developed for Pb-free assembly differ significantly from standard OSPs, approximately 290°C versus 260°C, respectively [7].

A thicker intermetallic compound (IMC) will be created after Pb-free soldering in comparison to that created with eutectic SnPb solder. The thicker IMC is caused by the increased peak reflow temperatures (+20 to 30°C) and the higher tin content of Pb-free solders resulting in a higher rate

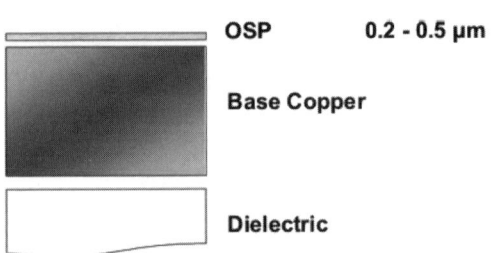

Fig. 9.5. OSP deposit showing relative layer thicknesses

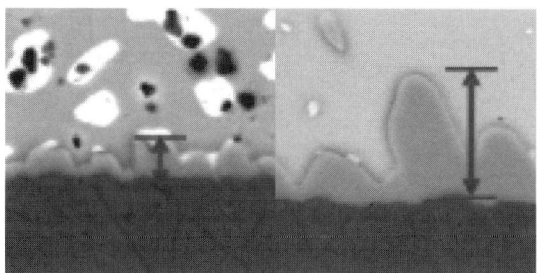

Fig. 9.6. Comparison of Cu-Sn IMC formed on OSP-treated surface with eutectic Sn37Pb solder, left, approx 2μm maximum thickness, and Sn3.0Ag0.5Cu Pb-free solder alloy, right, approx 5μm maximum thickness (Source: Atotech)

of dissolution of copper. Of course, this concern is valid for all finishes that form a Cu-Sn IMC, including immersion silver, immersion tin and direct immersion gold. HASL is also affected by this problem but to a greater extent due to its thermal excursion. Figure 9.6 demonstrates the difference in thickness of the Cu-Sn IMC formed with the two solders.

Process Description

Table 9.4 presents a typical process sequence for depositing the organic coating to a PWB.

Table 9.4. OSP – Typical Process Sequence (horizontal mode)

Process Step	Treatment Temperature (°C)	Treatment Time	Function
Clean	35-40	50-70 sec	Prepares copper surface for uniform treatment in subsequent microetch
Rinse	RT		
Microetch	20-25	50-70 sec	Provides defined roughness by exposing the copper grain structure
Rinse	RT		
Pre-coat (optional)	20-25	50-70 sec	Prepares the conditioned surface for deposition of the organic coating
OSP Coating	35-45	50-70 sec	Selectively deposits the organic coating on the copper surface
Rinse	RT		
DI Rinse	25-35		Optional rinse for reduced surface contamination
Dry			

Because of the relatively short treatment times, OSPs are typically applied using horizontal conveyorized equipment. The operation usually includes pretreatment of the copper surface (cleaning and microetching) prior to application of the OSP.

Key Issues - OSP

With the increased use of surface mount devices on PWBs, OSPs have become the most widely adopted method of surface finishing. Generally, OSPs are low-cost, simple to apply and offer excellent planarity. New formulations have been developed for Pb-free applications. However, there are still some inherent difficulties associated with these finishes. The coating is essentially invisible and therefore offers no color contrast for optical inspection. In addition, despite the new product developments, some concern still exists regarding the ability of the coatings to withstand multiple soldering operations, particularly when there is a subsequent wave soldering step.

Concern also exists regarding the spread characteristics of Pb-free solder pastes on OSP, particularly after one or more reflow operations. Recent testing concluded that Pb-free solder does not spread well on OSP-coated surfaces and wets the surface only where the paste is printed [8]. Reduced solder spreading and flow can lead to potential problems for both reflow soldering (in the form of incomplete coverage of pads) and wave soldering (in the form of insufficient hole-filling).

Performance of in-circuit testing (ICT) directly on the OSP-treated copper surface is not recommended. Problems may occur during ICT with the "bed of nails" fixture contact since the probe must penetrate the thermally hardened OSP or the oxidized copper surface in the case where the OSP has been removed. Overcoming this issue with increased probe force or change in pin design can result in increased risk of damaging the underlying copper, possibly leading to internal shorts. Normally, solder paste must first be reflowed on pads for ICT, but coverage of the test point is critical. As such, the test pad openings in the solder paste stencil are typically increased to compensate for this situation [9].

Other concerns associated with OSP involve its ability to be reworked. If solder paste is printed incorrectly and must be removed and reprinted, some or all of the OSP coating can also be removed. An IPC technical specification for OSP (IPC-4555) is currently being developed.

Table 9.5 provides an overview of the primary benefits and concerns regarding the OSP finish.

Table 9.5. Organic Solderability Preservative (OSP) – Process Summary

Benefits	Concerns/Limitations
+ Simple process/suitable for horizontal application	– Ability to withstand multiple thermal excursion (i.e. >3)
+ Suitable for Pb-free soldering	– Storage conditions required to maintain solderability
+ Good planarity	– Visual inspection difficulties; lack of color contrast
+ Shelf life of 6-12 months (with proper storage)	– Surface is not usually suitable "as is" for ICT inspection
+ Relatively easy to rework prior to any assembly activities	– Not wire-bondable
+ Suitable for use with mixed metal technology (ENIG)	– May require N_2 atmosphere for reflow/wave soldering
+ Low-cost deposit	– Removal of solder paste misprints may remove coating
+ Environmentally friendly	

9.2.3 Immersion Silver

Like most other Pb-free finishes, immersion silver appeared in response to the need for more planar pad surfaces for SMT applications. As an "immersion" process, a displacement reaction is involved in which silver ions are exchanged for copper ions and deposited on the exposed surface. Immersion silver is classified as a protective finish. During soldering, the molten alloy wets and spreads on the surface of the silver coating. The silver is then dissolved into the molten solder. The silver from the coating forms silver-tin intermetallic in the solder (Ag3Sn) with the molten solder wetting and spreading to the underlying copper, which must be equally solderable or dewetting will occur [10].

In the 1950s, restrictions to the use of silver in PWBs were adopted by Underwriters Laboratory (UL), based primarily on experience with electrochemical migration and dendrite formation with electroplated silver deposits. Electromigration is the tendency for metals to oxidize at the anode and re-deposit at the cathode in the presence of a liquid and a voltage potential. The re-deposited metal grows into a filament between the two electrodes, resulting in a potential short circuit [5]. However, extensive investigations were performed and ultimately led to a UL exemption for immersion silver, allowing its use in PWB fabrication without additional test requirements [11].

IPC has issued a technical specification (IPC-4553) for use of immersion silver as a surface finish for printed circuit boards [12].

Deposit Characteristics

Depending on the supplier, the deposit consists of a thin coating of silver, which may be deposited with a very small amount of organic material to reduce tarnish of the silver. The organic may be included either as an ingredient within the silver bath or applied in a subsequent step. For multiple soldering purposes, a minimum thickness of 0.15 μm (6 μin) is usually required, although the deposit usually ranges from 0.20 μm to 0.30 μm (8-12 μin), as illustrated in Figure 9.7. Any need for thicker silver usually involves wire-bonding applications [10].

Similar to OSP, with immersion silver as a surface finish an increase in IMC thickness will be realized with Pb-free solder in comparison to eutectic SnPb solder, as shown in Figure 9.8.

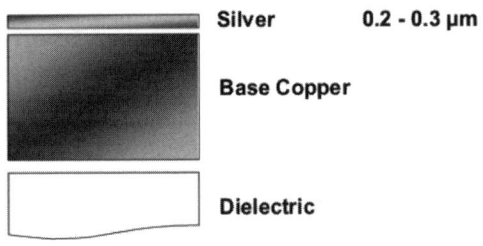

Fig. 9.7. Immersion silver deposit showing the relative layer thicknesses

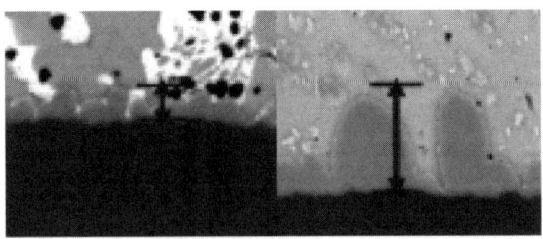

Fig. 9.8. Comparison of Cu-Sn IMC formed on immersion silver surface with eutectic Sn37Pb solder, left, approx 2μm maximum thickness, and Sn3.0Ag0.5Cu Pb-free solder alloy, right, approx 5μm maximum thickness (Source: Atotech)

Process Description

Immersion silver is a low-temperature process that can be applied to the PWB in either a vertical (dip) or horizontal (conveyorized) mode. Because treatment times are relatively short, the total process typically requires 10-15 minutes, depending on the proprietary process selected. Table 9.6 presents an example of a typical immersion silver process in the horizontal mode. There are several suppliers in the market and any variations to this process normally involve the use of an additional acid dip or "wetter" step preceding the immersion silver step or an optional post-dip step for improved anti-tarnish control.

Key Issues - Immersion Silver

Immersion silver is now used in many applications, including computers, consumer and automotive electronics. As a PWB finish, silver offers a simple, low-cost method of providing a highly planar surface that, with proper storage conditions, protects the underlying circuitry and is suitable for Pb-free soldering applications. Previous UL restrictions on immersion silver as a PWB finish have been removed. Still, the unsoldered silver coating remains a concern. When the board is heated without solder, the

Table 9.6. Immersion Silver – Typical Process Sequence (horizontal mode)

Process Step	Treatment Temperature (°C)	Treatment Time	Function
Acid Clean	35-60	50-70 sec	Prepares copper surface for uniform treatment in subsequent microetch step
DI Rinse	RT		
Microetch	30-40	60-80 sec	Provides defined roughness by exposing the copper grain structure
Rinse	RT		
Pre-Dip	40-50	50-70 sec	Prepares the conditioned surface for displacement
Immersion Silver	40-55	2-4 min	Deposits a fine, dense, uniform layer of silver by displacement of copper ions on the surface
Rinse	RT		
Post Dip (optional)	35-60	50-70 sec	Optional step to further reduce tarnishing and improve ionic cleanliness
DI Rinse Dry	25-35		

silver is known to tarnish, which can result in discolored pads that may cause solderability difficulties. Also, test pads (for ICT) should be printed with solder paste to avoid potential problems of exposed silver with dendritic growth or galvanic reaction, as shown in Figure 9.9. Similarly, due to the nature of the silver finish, shelf life of the product is a concern if the PWB is not properly protected and stored from corrosive airborne contaminants, such as sulfur and chloride.

To avoid such staining and tarnishing, proper handling and storage procedures must be followed to preserve the solderability properties of the immersion silver coating. Recent investigations have determined that the immersion silver deposit can migrate in harsh industrial environments (containing atmospheric sulfur and chlorides) or in humid environments in the presence of small amounts of chlorides, which are common in many field applications [13].

Recently, much attention has been focused on the issue of planar "microvoids" or "champagne voids". This type of voiding should not be confused with process related voids, which typically occur with any surface finish. In the case of planar voids, these tiny voids form at the interface between the pad and the solder, as seen in Figure 9.10. An X-ray examination of the solder joint, as shown in Figure 9.11, provides a more dramatic comparison of such microvoids and the more traditional process voids.

Fig. 9.9. Example of silver dendritic growth in extreme harsh environment test conditions (Source: Rockwell Automation)

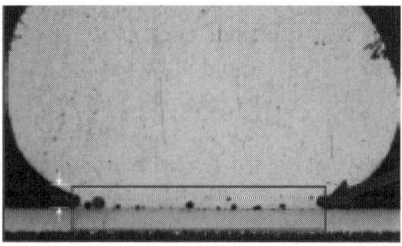

Fig. 9.10. Cross section of solder joint showing microvoiding at pad-solder interface (Source: Intel)

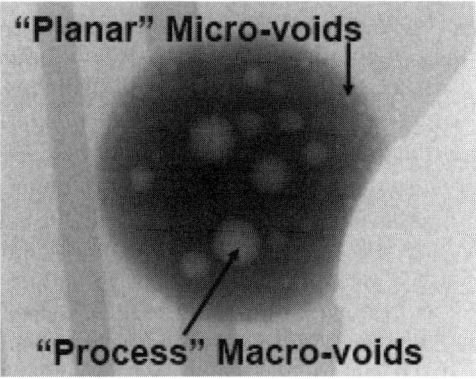

Fig. 9.11. X-ray of solder joint differentiating planar microvoids and process voids (Source: Intel)

There are numerous theories regarding the causes of this microvoiding. Some of the more commonly recognized contributing factors include:

- Out-gassing of the organic components within the immersion silver deposit
- Solder mask contamination on the copper
- Incompatible flux and/or solder paste
- Thick silver deposits ($>0.5\mu m(20uinch)$)
- Overly aggressive solder mask pretreatment causing significant copper roughness
- Incomplete microetching which alters the surface finish grain structure
- Decomposition of the immersion silver bath with aging
- Excessive plating deposition rate leading to 'copper caves' under the surface of the silver coating which are sites for voiding

Several theories for the reduction or elimination of planar microvoids have included [14]:

- Using a thinner immersion silver deposit
- Reducing the roughness of the silver deposit through slower plating and refining grain structure
- Reducing the amount of co-deposited organic

Table 9.7 provides an overview of the primary benefits and concerns regarding the immersion silver finish.

Table 9.7. Immersion Silver– Process Summary

Benefits	Concerns/Limitations
+ Simple process/suitable for horizontal application	– Tarnishing or staining (usually caused by reaction with airborne oxygen, chlorine or sulfur)
+ Suitable for Pb-free soldering	– Storage and handling conditions required to maintain solderability
+ Good planarity	– Planar microvoiding at the pad-solder interface
+ Shelf life > 12 months (with proper handling, storage conditions and deposit thickness)	– Potential for electromigration on exposed silver pads
+ Suitable for fine-pitch applications	– Potential for galvanic reaction with copper under solder mask
+ Relatively low-cost deposit	– Copper dissolution with additional soldering operations or high-temperature applications
+ Aluminum wire bondable	– Corrosion concerns for press-fit applications

9.2.4 Immersion Tin

The immersion tin process utilizes a displacement reaction that substitutes tin ions for copper ions to create a dense tin deposit on the exposed copper surface. Normally, the electropotential of copper is higher than that of tin. However, in the immersion tin process, thiourea is present to reduce the electropotential of copper such that the replacement reaction is driven in the desired direction.

As a surface finish, immersion tin is applicable for soldering, press fit connections and zero-insertion-force edge connections. Currently, IPC is developing a technical specification for use of immersion tin as a surface finish for printed wiring boards [15].

Deposit Characteristics

As a result of diffusion of copper through the tin deposit, a Cu-Sn inter-metallic compound (IMC) is created in the form of Cu_3Sn and Cu_6Sn_5. The growth of this IMC can impact both the shelf life and soldering perform-ance of the finish. As the thickness of the IMC expands to the surface, if all the pure tin is converted into Cu-Sn IMC no pure tin is left as a solderable layer and the wetting behavior will be adversely affected [15]. Figure 9.12 shows the relative thickness of the immersion tin deposit.

As shown in Figure 9.13, a thicker Sn-Cu IMC is known to occur with the use of Pb-free solder alloys, in comparison to that for eutectic SnPb solder. In particular, for multiple soldering processes it is essential to have a pure tin layer covering the Sn/Cu IMC before entering the final soldering process. The tin thickness is directly related to this IMC formation and typically ranges from 0.8 μm to 1.2 μm (32-48 μin). IPC draft specifications require a minimum thickness of 1.0 μm. For Pb-free soldering applications, a minimum tin thickness of 1.0 μm (40 μin) is typically recommended as a result of the increased IMC formation. The immersion tin deposit thickness is primarily influenced by process treatment time and/or temperature and is typically measured by X-ray fluorescence (XRF).

Process Description

As a PWB surface finish, immersion tin can be deposited in either a vertical (dip) or horizontal (conveyorized) mode. Because of the required tin

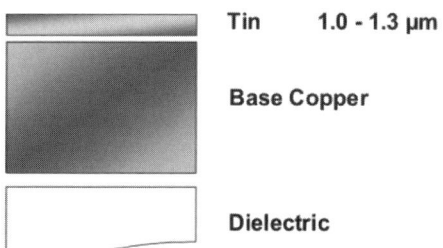

	Tin **1.0 - 1.3 μm**
	Base Copper
	Dielectric

Fig. 9.12. Immersion tin deposit showing relative layer thicknesses

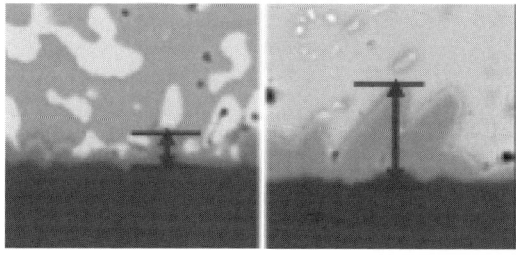

Fig. 9.13. Comparison of Cu-Sn IMC formed on immersion tin surface with eutectic Sn37Pb solder, left, approx 1.5μm maximum thickness, and Sn3.0Ag0.5Cu Pb-free solder alloy, right, approx 4μm maximum thickness (Source: Atotech)

thickness, the total treatment time can range from 15-20 minutes for horizontal application and 20 to 30 minutes for vertical application.

The functioning of the immersion tin solution can be influenced by several factors directly related to the process. For example, process temperature and time have a very strong influence on the rate of tin deposition and, as such, must be controlled within relatively narrow ranges. In addition, if not properly controlled, the concentration of Sn4+ can increase in the immersion tin solution to the point that it precipitates, sometimes leading to process problems in the form of reduced tin deposit thickness. As such, precautions are normally required to remove the Sn4+ from the process solution by periodic dilution/replacement or preferably by continuous treatment to convert Sn4+ to Sn2+. Also, as production continues, the copper concentration in the immersion tin solution will naturally increase. Although the appearance should not be adversely affected, the copper concentration must usually be maintained below 7 g/l for thickness control.

Table 9.8 presents an example of an immersion tin process sequence in the horizontal mode, including key parameters.

Some variations to this model may include a post-treatment step for improved control of ionic contamination. Immersion tin processes that follow this basic schematic are available commercially from several suppliers.

Table 9.8. Immersion Tin – Typical Process Sequence (horizontal mode)

Process Step	Treatment Temperature (°C)	Treatment Time	Function
Acid Clean	35-60	50-70 sec	Remove copper oxide and solder mask developing residue
Rinse	RT		
Microetch	25-35	70-90 sec	Provides defined roughness by exposing the copper grain structure
Rinse	RT		
Pre Dip/ Conditioner	20-30	50-70 sec	Conditions copper surface to control the ion replacement reaction
Immersion Tin	65-75	8-12 min	Deposit a fine, dense, uniform layer of tin by displacement of copper ions on the surface
Post Dip			Removal of tin salts (surface contamination)
Rinse	RT		
Dry			

Fig. 9.14. Production line for application of immersion tin (Source: Atotech)

Figure 9.14 shows an example of a production system in horizontal mode for the application of immersion tin as a PWB surface finish.

Immersion Tin Key Issues

Immersion tin has gained acceptance as a surface finish for numerous applications, including automotive, computer and consumer electronics. The process is a relatively simple, cost-effective method to provide a planar deposit that is well suited to the tin-rich composition of the Pb-free solders. Immersion tin is also used for compliant pin connectors, which typically utilize tin-plated surfaces. However, like other finishes, potential problems also exist with immersion tin. As previously described, the tin deposit must be thick enough to contain the growth of the Cu_6Sn_5 IMC, which can be affected by storage time, storage conditions and number of thermal excursions. Therefore, the thickness of the immersion tin deposit must be carefully controlled. If less than 1.0 µm (40 µin) of tin is applied, there is a potential that the Cu-Sn IMC growth will be such that insufficient pure tin will exist to withstand multiple Pb-free soldering operations. As such, the draft IPC specifications for immersion tin (IPC-4554) stipulate a minimum deposit of 1.0 µm [15].

The subject of tin whiskers in electronic applications has been well documented since its discovery in the early 1950s. Tin whiskers are hairline, mono-crystals that are theorized to develop and grow as a result of internal stress (mechanical, thermal, and other forms). An example of a tin whisker is shown in Figure 9.15. Immediately after plating, immersion tin layers exhibit tensile stress. However, as the Cu_6Sn_5 IMC grows, internal compressive stresses are induced. its growth is unchecked, the whisker can be the cause of a short circuit. Under certain conditions the immersion tin deposit has been shown to be susceptible to whisker growth. The possibility of whisker formation also exists as a direct consequence of stresses induced on the immersion

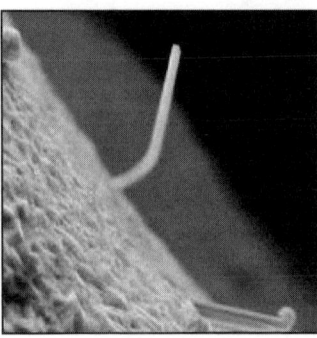

Fig. 9.15. Tin whisker at 6000X magnification (Source: Atotech)

tin deposit as a result of compliant pin (i.e. "press fit") insertions [15]. Whiskers up to 150µm in length have been documented in PWB vias, although much shorter whiskers have been reported on the edges of surface mount pads [16]. Thermal treatment (at conditions similar to reflow soldering) can reduce or even prevent whisker growth by lowering the internal stress by causing movement of tin atoms that reduces the lattice defects. However, this method is not typically employed as a preventive measure because it also accelerates the IMC growth, thus limiting the shelf life. Some suppliers have minimized or prevented whisker growth with the addition of a (metal or organic) ingredient to the immersion tin reaction. IPC has taken the position that it is the responsibility of the end user to determine the impact of potential whisker growth on long-term reliability.

An overview of the benefits and concerns of the immersion tin finish is summarized in Table 9.9.

Table 9.9. Immersion Tin– Process Summary

Benefits	Concerns/Limitations
+ Simple process/suitable for horizontal application	– Control for tin whisker growth
+ Shelf life > 12 months (with proper handling, storage and deposit thickness)	–Thickness of pure tin in relation to IMC thickness
+ Suitable for Pb-free soldering	– Storage and handling conditions required to maintain solderability
+ Good planarity	– Solder mask compatibility
+ Suitable for fine-pitch applications	– Not suitable for contact switching
+ Relatively low-cost deposit	– Not suitable for wire bonding

9.2.5 Electroless Nickel/Immersion Gold (ENIG)

Since the early 1990s, one of the more versatile PWB surface finishes has been electroless nickel/immersion gold (ENIG). As the name implies, ENIG is a dual-metal deposition process. Following pretreatment and activation, electroless nickel is deposited on the exposed, activated copper surface. A very thin layer of gold is then deposited via an immersion reaction which displaces nickel on the surface. The primary purpose of the immersion gold layer is to protect the nickel surface from oxidation and passivation until it can be soldered. The nickel layer forms a barrier between the gold and underlying copper, creating a Ni-Sn IMC. With sufficient nickel thickness, the copper is protected from dissolution by the solder. Such dissolution is even more aggressive in the case of Pb-free solder alloys. The nickel layer also adds strength to plated-through holes and vias, while increasing wear resistance through its inherent hardness [17].

Deposit Characteristics

Electroless nickel is technically a co-deposition process whereby nickel is co-deposited with phosphorus on the copper surface at a thickness of 3.0 μm to 6.0 μm (120-240 μin). Depending on the process, the phosphorus content can vary from approximately 5-12 percent by weight. Higher concentrations of phosphorus have been reported to improve the corrosion resistance of the nickel deposit, an advantage for harsh environment applications [18]. The immersion gold layer, at 99.99% purity, is deposited on the nickel to a thickness of approximately 0.05 μm to 0.10 μm (2-4 μin). At such thickness, the immersion gold deposit does provide good corrosion resistance and prevents oxidation of the underlying nickel. Figure 9.16 illustrates the relative thicknesses of the deposits.

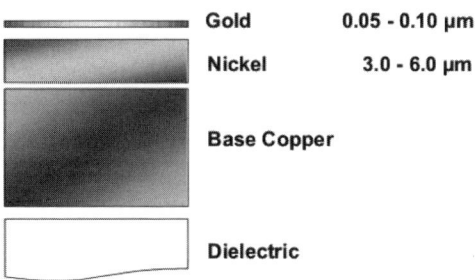

Gold	0.05 - 0.10 μm
Nickel	3.0 - 6.0 μm
Base Copper	
Dielectric	

Fig. 9.16. ENIG deposit showing relative layer thicknesses

Process Description

Table 9.10 presents a typical sequence for the ENIG process, including treatment temperatures and times.

Table 9.10. Electroless Nickel/Immersion Gold (ENIG) – Typical Process Sequence

Process Step	Treatment Temperature (°C)	Treatment Time	Function
Clean	35-45	3-6 min	Prepares copper surface for uniform treatment in subsequent microetch
Rinse	RT	60 sec	
Microetch	25-35	1-2 min	Provides defined roughness by exposing the copper grain structure
Rinse	RT	60 sec	
Acid Dip	RT	3-5 min	Prepares copper surface for subsequent conditioning and protects conditioner from drag-in
Activate/ Catalyst	20-25	1-3 min	Provides an extremely thin, active seed layer for autocatalytic deposition of nickel
Rinse	RT	60 sec	
Electroless Nickel	80-90	15-30 min	Autocatalytic deposition of nickel onto the activated copper surface
Rinse	RT	60 sec	
Immersion Gold	80-90	7-14 min	Displacement of nickel ions with gold
Rinse	RT	60 sec	
Dry			

Electroless nickel is an autocatalytic process, meaning that nickel is continuously deposited onto the pretreated and exposed copper surface through an oxidation-reduction chemical reaction. Figure 9.17 shows an example of a PWB production system for the application of ENIG.

Fig. 9.17. Production line for vertical application of ENIG (Source: Atotech)

Both the electroless nickel and immersion gold segments of the process are high-temperature steps, with the nickel requiring treatment times up to 30 minutes. For this reason, use of a suitable solder mask and secondary image technology (SIT) photoresist is essential. Also, the gold deposition is an immersion step; requiring the removal of nickel ions for the deposition of gold to occur. The process must be controlled such that the immersion gold does not result in excessive corrosion or attack of the nickel.

Key Issues - ENIG

Because of its corrosion resistance, high planarity and excellent shelf life, ENIG has been employed for many years as a PWB surface finish. It has often been used in applications where the more traditional finishes (i.e. HASL or OSP) lack certain properties necessary for storage or functionality, such as contact switching or aluminum wire bonding.

Solder joint formation of nickel/gold surface finishes occurs through gold dissolution into the solder, allowing the formation of an IMC between the nickel and tin. It is well known that the formed Ni-Sn IMC influences the solder joint integrity. The dissolution rate of nickel into tin is slower than that of copper into tin. Therefore, in comparison, the growth rate of the IMC between Sn-based solders and nickel is relatively slow [5]. With eutectic solder, this results in a Ni_3Sn_4 intermetallic with a thickness of only 0.25-0.75 μm after a typical soldering process. With Pb-free solder employing a SnAgCu alloy, Ni_3Sn_4 and $[Cu,Ni]_6Sn_5$ IMC are formed.

Brittle materials have little plasticity and will fail under loading without stretching or changing shape. Young's Modulus indicates the relative brittleness of the individual compounds, where a higher value indicates a more brittle nature. In general, the Ni_3Sn_4 IMC is more brittle than Cu_6Sn_5 and Cu_3Sn IMCs. [19]. Figure 9.18 graphically compares the Young's Modulus of Elasticity for these isolated IMCs.

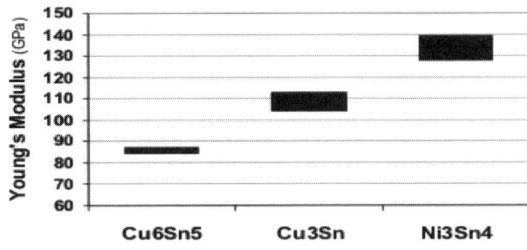

Fig. 9.18. Comparison of Young's Modulus for Cu-Sn and Ni-Sn IMCs (Source: Atotech)

It is also possible for solder joint problems to be related to gold embrittlement, caused by excessive gold dissolution during soldering, creating a brittle $AuSn_4$ IMC. However, this condition does not usually appear until the gold content in the solder joint approaches approximately three- to four-percent by weight for Sn37Pb solder, which should not occur if the immersion gold thickness is maintained within the recommended range [20]. Lead-free solders containing high amounts of tin may be more tolerant to gold content and the resultant formation of $AuSn_4$ IMC in comparison to Sn37Pb solder [21]. However, it should be noted that previous investigations have determined that after dissolving into the solder even thin layers of immersion gold were prone to re-deposition at the solder/nickel interface as $AuSn_4$ IMC [22].

One well known problem associated with ENIG is the "black pad" phenomenon (a.k.a. black nickel or black-line nickel), which can cause solder joint failures after assembly. The defective solder joint appears black or dark gray. Examination of the electroless nickel by SEM typically reveals "mud cracks" in the topography on the surface, as shown in Figure 9.19. Also, cross-sectioning of the affected area shows a "tooth decay" effect of corrosion of the nickel layer [23]. This condition is usually an indication of irregularities in the nickel deposit or excessive exposure in the immersion gold bath that results in nickel corrosion. In either case, the corrosion of the nickel layer is directly attributed to the immersion gold reaction. Therefore, it is essential that the proper nickel plating parameters (i.e. treatment time, temperature, pH, phosphorus content, bath metal turnovers, etc.) be maintained to ensure a deposit that will be compatible with the subsequent immersion gold.

Theories vary regarding the impact of the phosphorus content of the nickel on the performance of the finish. It was once thought that the black

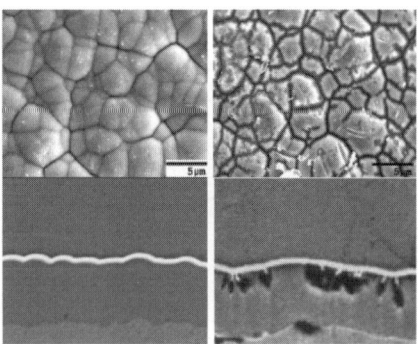

Fig. 9.19. Example of surface and cross-section views of normal nickel deposit, left, and nickel deposit with "black pad" effect, right (Source: Atotech)

Table 9.11. Electroless Nickel Immersion Gold (ENIG) – Process Summary

Benefits	Concerns/Limitations
+ Suitable for Pb-free soldering	– Cost of deposit
+ Good planarity	– Potential brittleness of nickel-tin IMC
+ Barrier layer (nickel) to stop dissolution of copper	– Corrosion of nickel by gold, leading to "black pad" issue
+ Good shelf life (>12 months) with less susceptibility to environment conditions	– Process operating window; requires tight control
+ Good surface for ICT probability	– Not suitable for gold wire bonding applications
+ Suitable for contact switching	– Relatively high level of maintenance
+ Suitable for aluminum wire bonding applications	– Excessive nickel plating (bridging) and/or skip plating (no nickel)

pad effect was directly related to enrichment of phosphorus near the interface between the nickel and solder. As such, logic proposed that lower phosphorus content in the electroless nickel would reduce the occurrence of black pad. However, more recent studies have shown that an increase in the phosphorus content to a range of 8-12% actually reduces the corrosion during the immersion gold process [18]. Other studies have also reported similar findings; however, there is some concern that phosphorus content above 10% may adversely affect the rate of the immersion gold process [24]. As a result, some suppliers of high-phosphorus ENIG processes have developed immersion gold baths tailored for such applications.

IPC has issued a technical specification (IPC-4552) for use of ENIG as a surface finish for printed circuit boards [25].

A summary of the benefits and concerns of the ENIG finish is presented in Table 9.11.

9.2.6 Electroless Nickel/Electroless Palladium/Immersion Gold (Ni/Pd/Au)

As electronic devices have become smaller and more complex, so have the assembly methodologies become more diverse. In addition to the demands caused by the higher temperatures of Pb-free soldering on increasingly smaller pads, certain assemblies may include a mix of soldered as well as wire-bonded components. The electroless nickel/electroless palladium/ immersion gold (Ni/Pd/Au) process evolved from the previously described ENIG process to meet the needs of PWBs requiring a higher level of solder joint reliability or requiring multiple assembly operations, such as soldering and wire bonding, as shown in Figure 9.20. It was determined that deposition of an electroless palladium layer between the electroless nickel and immersion

Fig. 9.20. The Ni/Pd/Au finish is suitable for soldering as well as gold and aluminum wire bonding (Source: Atotech)

gold layers improved the solder joint integrity of the ENIG deposit, particularly at the higher reflow temperatures required in Pb-free assembly [26]. Surface finishes of electroless palladium over electroless nickel have been previously described [5]. The addition of the immersion gold deposit on the palladium was determined to extend the product shelf life and improve soldering and wire bonding behavior.

Recent investigations have shown that the measured level of solder wetting of the Ni/Pd/Au surface throughout three reflow cycles in nitrogen atmosphere was at least two times greater than that achieved by Pb-free HASL, OSP, immersion silver or immersion tin [27]. The finish is also used for certain non-organic substrates, such as low-temperature co-fired ceramics (LTCC), used in harsh environments (automotive, aerospace, etc).

Deposit Characteristics

The Ni/Pd/Au surface finish is essentially an ENIG deposit with a high-purity palladium layer (99.99% Pd) sandwiched between the nickel and gold. Figure 9.21 shows the range in thickness of the respective deposits.

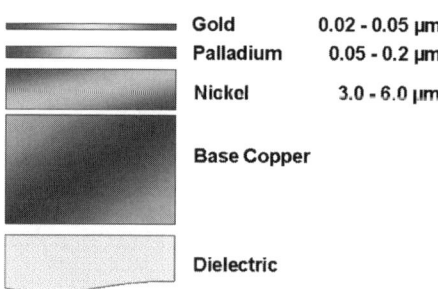

Gold	0.02 - 0.05 μm
Palladium	0.05 - 0.2 μm
Nickel	3.0 - 6.0 μm
Base Copper	
Dielectric	

Fig. 9.21. Ni/Pd/Au deposit showing relative layer thicknesses

The electroless nickel and immersion gold deposits are the same as those achieved with ENIG. The palladium layer is typically deposited as a pure metal or co-deposited with phosphorus. Because of the added palladium layer, a somewhat thinner gold deposit can be applied. It has been theorized that the palladium layer acts to delay the diffusion of the nickel into Pb-free solders, confirmed by a thinner Cu-Ni-Sn IMC in comparison to that produced with ENIG. [26]. In eutectic SnPb solder applications, it has been proposed that the palladium in the Ni-Sn IMC tends to form in clusters that are separated from the lead in the joint, thus creating a non-uniform IMC [28]. In one study, it was found that for both Ni/Pd/Au and ENIG a Cu-Ni-Sn IMC is formed, but it is thinner for Ni/Pd/Au than for ENIG, both before and after thermal conditioning. It was considered that the palladium layer inhibited the diffusion of nickel into the solder to give a thinner Cu-Ni-Sn IMC. In addition, growth rate of Cu-Ni-Sn IMC is slower for Ni/Pd/Au than the other case [29]. Figure 9.22 shows a comparison of the $(Cu,Ni)_6Sn_5$ IMC created on NiPdAu surface finish with SAC alloy solder after different reflow conditions [27]. In this case, the palladium and gold layers are very thin, both diffusing into the bulk solder.

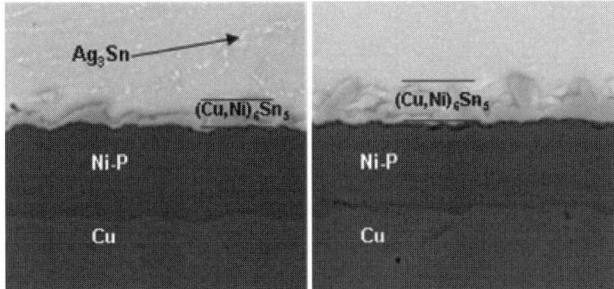

Fig. 9.22. Comparison of $(Cu,Ni)_6Sn_5$ IMC created on NiPdAu surface finish with Sn3Ag0.5Cu solder after 1 reflow, left, approx 1.7μm maximum thickness and after 5 reflows, right approx 3.8μm maximum thickness (Source: Atotech)

Process Description

Table 9.12 presents a typical process sequence for the application of the Ni/Pd/Au surface finish. The sequence is very similar to the previously described ENIG process, with the exception of the additional palladium deposition step. In this case, a thin layer of palladium is autocatalytically deposited onto the nickel surface.

Table 9.12. Electroless Nickel/Electroless Palladium/Immersion Gold (Ni/Pd/Au) – Typical Process Sequence

Process Step	Treatment Temperature (°C)	Treatment Time	Function
Clean	35-45	3-6 min	Prepares copper surface for uniform treatment in subsequent microetch step
Rinse	RT	60 sec	
Microetch	25-35	1-2 min	Provides defined roughness by exposing the copper grain structure
Rinse	RT	60 sec	
Acid Dip	RT	3-5 min	Prepares copper surface for subsequent conditioning and protects conditioner from drag-in
Activate/ Catalyst	20-25	1-3 min	Provides an extremely thin, active seed layer for autocatalytic deposition of nickel
Rinse	RT	60 sec	
Electroless Nickel	80-90	15-30 min	Autocatalytic deposition of nickel onto the activated copper surface
Rinse	RT	60 sec	
Electroless Palladium	50-70	5-10 min	Autocatalytic deposition of palladium onto the nickel surface
Rinse	RT		
Immersion Gold	80-85	7-14 min	Displacement of palladium ions with gold
Rinse	RT	60 sec	
Dry			

Key Issues – Ni/Pd/Au

As a surface finish Ni/Pd/Au has been available for more than ten years. However, its primary use has been associated with the finishing of lead frames or as a package substrate finish, where wire-bonding requirements are more common. With increased solder joint integrity demands associated with Pb-free processing, this finish is now receiving more attention for PWB applications. Its benefits are similar to those of ENIG with the added advantages of improved solder joint reliability and its suitability for gold wire bonding applications. However, the costs and complexity of the

ENIG process are even further increased with the added electroless palladium step. Furthermore, although the solder joint integrity is reportedly improved, the issue of the brittle nature of the Ni-Sn IMC is not completely eliminated. Previous investigations have shown that Sn-based solder joints are susceptible to a loss in ductility as the Pd loading in them is increased. These same examinations concluded that to ensure proper ductility of the solder joint the Pd thickness should not exceed 0.35 μm [5]. More recently, it was determined that the thickness of the palladium layer in the Ni-Pd-Au deposit has a significant effect on the fracture mode of eutectic SnPb BGA soldered joints. Optimum results were actually achieved with less than 0.2 μm of Pd and applying a thicker deposit resulted in increased brittle fracture [30]. This could be due to the formation of $PdSn_4$ intermetallic, which is brittle like $AuSn_4$ IMC. However, continuing examinations indicate that the Pd thickness may have less of an effect on the brittleness of Pb-free solder joints. In light of some unanswered questions, perhaps the most significant concern is the lack of experience with this process within the PWB industry.

Table 9.13 summarizes the benefits and concerns related to the Ni/Pd/Au surface finish.

Table 9.13. Electroless Nickel/Electroless Palladium/Immersion Gold (Ni/Pd/Au) – Process Summary

Benefits	Concerns/Limitations
+ Shelf life > 12 months	– Relatively high cost
+ Palladium layer reduces formation/ amount of brittle Ni-Sn IMC	– Relatively complex process (more process steps than ENIG)
+ Little or no incidence of "black pad" effect (associated with ENIG)	– Potential brittleness of nickel-tin IMC
+ Suitable for Pb-free soldering	– Potential for solder joint embrittlement due to excess Pd ($PdSn_4$ IMC)
+ Good planarity	– Process operating window; requires tight control
+ Barrier (nickel) to stop dissolution of copper	– Excessive nickel plating (bridging) and/or skip plating (no nickel)
+ Suitable for gold and aluminum wire bonding applications	– Relatively unknown process with little PWB history for this surface finish
+ Good surface for ICT probability	
+ Suitable for contact switching applications	

9.2.7 Electrolytic Nickel/Electrolytic Gold

For PWB applications, most electrolytic nickel/electrolytic gold deposits are classified as either "hard" or "soft" gold. Hard gold is typically used for connection purposes that can receive heavy wear or require repeated insertion or removal. The most common example would be edge connectors, tabs or "fingers" along the side of the PWB. These are usually plated in equipment designated for such purposes. Hard gold is often used for compliance with military standards [31]. In contrast, soft gold or "pure" gold is primarily used where gold wire bonding of an IC is required. Although gold wire bonding is more commonly performed on IC package substrates, chip-on-board (COB) technology may require gold wire bonding directly on the PWB. As with ENIG, the nickel deposit serves as a barrier between the copper and gold layers.

Plating of electrolytic nickel and electrolytic gold requires an applied current and a bus connection incorporated into the circuit design. The current increases the deposition rate and provides a more dense coating. However, as an electrolytic process the deposit thickness is more variable, caused by current density and circuit design issues.

Deposit Characteristics

For hard gold applications, the electroplated nickel will greatly increase the wear resistance of the gold-plated surface. As in ENIG, the nickel acts as a barrier to prevent diffusion of the underlying copper into the gold. Electrolytic nickel is typically deposited within a thickness range of 2.5-7.0 μm, depending on the application or performance needs.

Hard gold processes typically provide a deposit of about 99.7% purity. Electrolytic gold deposit thickness requirements can vary significantly with application and solder alloy. Edge connector applications may typically require a thickness range of 1.25 μm (50 μin) to 2.5 μm (100 μin) [31]. However, for soldering purposes, a gold thickness of 0.5 μm (20 μin) is typically deposited. However, it must also be noted that for eutectic solder applications, the localized concentrations of gold in the solder joint must be limited to 3 to 4 weight percent. Above this concentration, a condition of gold embrittlement can be caused by creation of the $AuSn_4$ IMC, which can potentially lead to crack initiation or propagation sites in the solder

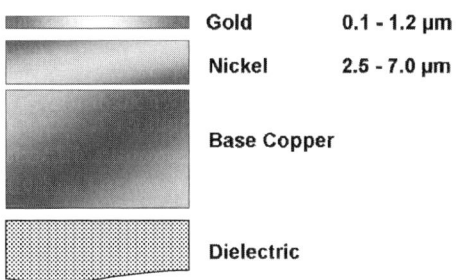

	Gold	0.1 - 1.2 µm
	Nickel	2.5 - 7.0 µm
	Base Copper	
	Dielectric	

Fig. 9.23. Electrolytic nickel/electrolytic gold deposit showing relative layer thicknesses

joint. By comparison, for high tin Pb-free solder alloys (comprised of Sn-Ag or Sn-Ag-Cu) a higher concentration of gold may be tolerated in the solder joint before signs of embrittlement due to $AuSn_4$ IMC occur [21]. As such, to prevent gold embrittlement the gold thickness must be carefully specified and controlled in relation to solder volume to pad dimensions.

For other applications not involving wear but requiring corrosion resistance and soldering, the necessary hard gold deposit thickness will typically be 0.10 µm to 0.50 µm (4-20 µin). For wear resistance applications (i.e. edge connectors) a hard gold deposit thickness range of 0.8 µm to 1.2 µm (32-48 µin) is typically required. For applications such as wire bonding, soft gold is normally specified. The required thickness of this high-purity (>99.99%) gold may be 0.5 µm to 1.0 µm (20-40 µin). Figure 9.23 shows the range of deposit thickness required for the various finish applications.

Process Description

Electroplated nickel/gold is typically deposited after acid copper plating but before etch and soldermask application and thus can also be used as an etch resist for those features affected. Electrolytic nickel is normally applied in a process bath consisting of nickel sulfamate or nickel sulfate. Electrolytic gold provides a coating only on the top surfaces of all exposed nickel traces and features. Therefore, after final etch operations, the trace and feature sidewalls (i.e. Cu and Ni) are exposed [17].

Table 9.14 presents a typical sequence for an electrolytic nickel/electrolytic gold process, including treatment temperatures and times.

Table 9.14. Electrolytic Nickel/Electrolytic Gold – Typical Process Sequence

Process Step	Treatment Temperature (°C)	Treatment Time	Function
Acid clean	35-45	3-6 min	Prepares copper surface for uniform treatment in subsequent microetch
Rinse	RT	60 sec	
Microetch	25-35	1-2 min	Provides defined roughness by exposing the copper grain structure
Rinse	RT	60 sec	
Acid Dip	RT	3-5 min	Prepares copper surface for subsequent treatment step and prevents drag-in of contaminants
Rinse			
Electrolytic Nickel	50-60	Thickness dependent	Electrochemical deposition of nickel onto exposed copper surface
Rinse			
Gold strike	40-60	Thickness dependent	Electrochemical deposition of gold onto exposed nickel surface
Electrolytic Gold	35-50 (hard Au) 60-70 (soft Au)	Thickness dependent	
Rinse Dry	RT	60 sec	

Key Issues - Electrolytic Nickel/Electrolytic Gold

The electrolytic nickel/electrolytic gold process is certainly not a mainstream surface finish for PWBs, but does have its applications. Selection of this process usually indicates some special board feature, assembly requirement or end use specification that precludes the use of the more common finishes, previously described.

Because of the relatively thick gold layer, electrolytic Ni/Au is probably the most expensive surface finish applied to PWBs. However, in this case cost does not necessarily equate to technology demand. In some cases, its use is considered as an alternative to OSP in applications such as lead-free wave soldering, particularly in thick PWBs where copper dissolution from through-holes can be reduced because of the nickel barrier layer, while also achieving improved hole-fill. Furthermore, unlike ENIG, the process is not prone to develop black pad because the gold deposition occurs by electrolytic reaction, which does not attack the nickel layer in comparison to the immersion reaction in ENIG. However, the Ni-Sn IMC formation is still a concern because of its inherent brittleness.

Table 9.15. Electrolytic Nickel/Electrolytic Gold – Process Summary

Benefits	Concerns/Limitations
+ Long shelf life (> 12 months)	– High cost of deposit
+ Suitable for Pb-free soldering	– Requires electrical continuity within PWB (i.e. bussing)
+ Barrier (nickel) to stop dissolution of copper	– Excess gold in solder joint can cause embrittlement
+ Suitable for contact switching applications	– Potential brittleness of nickel-tin IMC
+ Suitable for gold wire bonding applications	– Plating thickness variations for both surface and through holes
+ Suitable for compliant pin (press-fit) connections	– Exposed copper on sidewalls of traces and features
+ Good surface for ICT probability	
+ No black pad issue (concern for ENIG)	

In comparison to "chemical" processes (immersion or electroless), the technology level achievable with an electrolytic deposition is somewhat limited in terms of surface distribution and wiring/interconnect density. The variations in deposit thickness can be significant. In addition, the electroplated metal also has a higher inherent stress level compared to electroless or immersion deposits. Because electroplating proceeds more rapidly and is not self-limiting, both nickel and gold thickness must be closely controlled. Finally, valuable PWB "real estate" must be surrendered for the bussing required for electrolytic processing.

From the standpoint of soldering, there may be sufficient gold available to create embrittlement in the solder joint by producing a sufficient amount of $AuSn_4$ IMC. Embrittlement typically occurs when the solder joint contains gold at more than three to four weight percent. The presence of $AuSn_4$ IMC will reduce the strength of the bond between the joint and the base metal and can result in the entire joint becoming detached [32]. Also, the underlying nickel will still form a Ni-Sn IMC, such as Ni_3Sn_4, which is more brittle than Cu-Sn IMCs, as previously noted.

Table 9.15 provides a summary of the primary benefits and concerns regarding the electrolytic nickel/electrolytic gold finish.

9.2.8 Electroless Nickel/Immersion Gold/Electroless Gold (ENIGEG)

For applications requiring gold wire bonding, an alternative approach to electrolytic nickel/electrolytic gold would be the electroless nickel/immersion

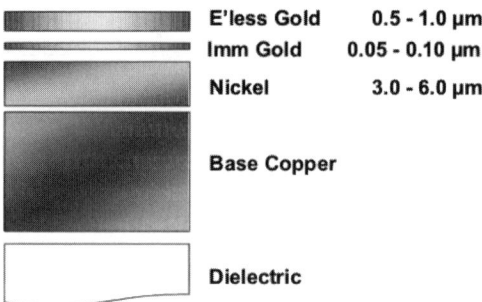

	E'less Gold	0.5 - 1.0 μm
	Imm Gold	0.05 - 0.10 μm
	Nickel	3.0 - 6.0 μm
	Base Copper	
	Dielectric	

Fig. 9.24. Electroless nickel/immersion gold/electroless gold deposit showing relative layer thicknesses

gold/electroless gold (ENIGEG) process. Often, the electroless gold will be applied selectively to the wire bond pads only, which does require a second photoresist imaging process. When this is not possible or practical, the surface finish must provide for both wire bonding and solderability functions, but a concern still exists regarding the gold thickness/embrittlement issue.

Electroless nickel/electroless gold does offer some distinct advantages over the previously described fully electrolytic process. Since it is an electroless process, it requires no electrical buss within the PWB and can be applied after copper circuitization.

Deposit Characteristics

The electroless nickel and immersion gold deposits are similar to those applied in ENIG, approximately 3.0-6.0 μm (120-240 μin) of nickel and 0.05 to 0.10 μm (2-4 μin) of gold. After ENIG, electroless gold is subsequently applied to build the total gold thickness to as much as 1.0 μm (40 μin), using the immersion gold as the seed layer for autocatalytic gold deposition. However, for most gold wire bonding applications, a nominal thickness of 0.5 μm (20 μin) is often required. Figure 9.24 illustrates the relative thicknesses of the deposits.

Process Description

Because an immersion gold step is required prior to the deposition of electroless gold, the process is essentially the same as the standard ENIG process with the subsequent deposition of electroless gold. A summary of the

Table 9.16. Electroless Nickel/Immersion Gold/Electroless Gold (ENIGEG) – Typical Process Sequence

Process Step	Treatment Temperature (°C)	Treatment Time (sec)	Function
Clean	35-45	3-6 min	Prepares copper surface for uniform treatment in subsequent microetch
Rinse	RT	60 sec	
Microetch	25-35	1-2 min	Provides fine-grained, defined roughness by exposing the copper crystal structure
Rinse	RT	60 sec	
Acid Dip	RT	3-5 min	Prepares copper surface for subsequent conditioning and protects conditioner from drag-in
Activate/ Catalyst	20-25	1-3 min	Provides an extremely thin, active seed layer for autocatalytic deposition of nickel
Rinse	RT	60 sec	
Electroless Nickel	80-90	15-30 min	Autocatalytic deposition of nickel onto the activated copper surface
Rinse	RT	60 sec	Removes process chemicals from previous step
Immersion Gold	80-90	8-14 min	Displacement of nickel ions with gold
Rinse	RT	60 sec	
Electroless Gold	85-95	40-50 min	Autocatalytic deposition of gold onto the immersion gold (seed) layer
Rinse	RT	60 sec	
Dry			

sequence and key parameters for the electroless nickel/electroless gold process is shown in Table 9.16.

Key Issues - Electroless Nickel/Immersion Gold/Electroless Gold

The ENIGEG surface finish is not typically specified for applications involving soldering alone. Typically, the driving force is another assembly need, such as gold wire bonding. However, the surface finish must also meet the soldering requirements of other components. In terms of applying the finish, ENIGEG poses many of the same issues as the previously described ENIG process. Thus, the potential for nickel attack by immersion gold and the accompanying black pad effect must be recognized. Also, similar to ENIG, creation of the somewhat brittle nickel-tin IMC remains a concern. With the thicker gold deposit, embrittlement of the solder joint

Table 9.17. Electroless Nickel/Electroless Gold – Process Summary

Benefits	Concerns/Limitations
+ Suitable for Pb-free soldering	– High cost of deposit
+ Good planarity	– Potential brittleness of nickel-tin IMC
+ Barrier layer (nickel) to stop dissolution of copper	– Corrosion of nickel by immersion gold, leading to "black pad" issue
+ Long shelf life (>12 months)	– Excess gold in solder joint can cause embrittlement
+ Good surface for ICT probability	– Process operating window; requires tight control
+ Suitable for contact switching applications	– Solder mask compatibility
+ Suitable for gold wire bonding applications	– Excessive nickel plating (bridging) and/or skip plating (no nickel)
+ No need for electrical continuity within PWB (i.e. bussing)	

(as a result of formation of the AuSn$_4$ IMC) must also be considered. Furthermore, the added electroless gold step exposes the PWB to an extended plating time (45 min) at high temperature that can result in attack of the solder mask. Because of the required thickness of the gold deposit, the costs associated with the process are similar to those of electrolytic nickel/electrolytic gold. In comparison to that process, however, there are some benefits to this version, including better planarity and the lack of requirement for electrical bussing.

Table 9.17 provides a summary of the primary benefits and concerns regarding the electroless nickel/electroless gold finish.

9.2.9 Immersion Gold (over copper)

Commonly referred to in the industry as "direct gold" or "direct immersion gold" (DIG), this relatively new process is actually a hybrid of the immersion gold and electroless gold reactions, employing a mix of displacement and autocatalytic reactions. It has been reported that the autocatalytic portion of the reaction contributes at least 80-percent of the total deposit [33]. Immersion gold can be considered as an alternative for applications similar to those served by OSP, immersion silver and immersion tin finishes.

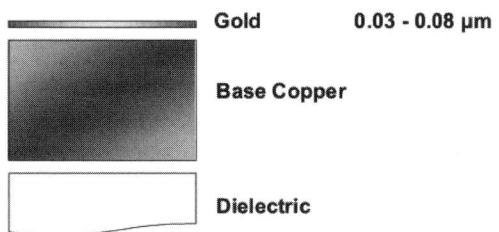

Fig. 9.25. Immersion gold deposit showing relative layer thicknesses

Deposit Characteristics

As a noble metal, gold should not oxidize and, therefore, a very thin layer will provide protection of the underlying copper. As shown in Figure 9.25, the required thickness of the gold coating is typically 0.03 to 0.08 μm. Figure 9.26 shows the sequential change in surface topography during a 60-minute process treatment in immersion gold.

Figure 9.27 shows the effect of high-temperature conditioning on the development of the copper/tin IMC, when using a Pb-free solder (SAC305) with an immersion gold surface finish. Similar to ENIG, because the gold layer is so thin the issue of embrittlement (from AuSn4 IMC) should not occur if the immersion gold thickness is maintained within the recommended range. As shown, two distinct IMCs are formed, Cu3Sn and Cu6Sn5 [34]. Because of the very thin nature of the deposit, there is a potential for copper migration through the gold. Therefore, it is necessary for PWBs with this finish to be assembled within four months of its application [35]. Such a requirement would appear to limit any widespread use of the finish.

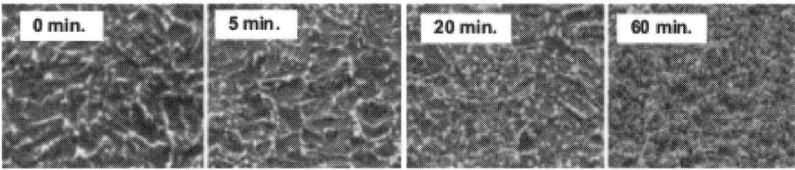

Fig. 9.26. Change in surface topography during deposition of direct immersion gold (Source: Uyemura)

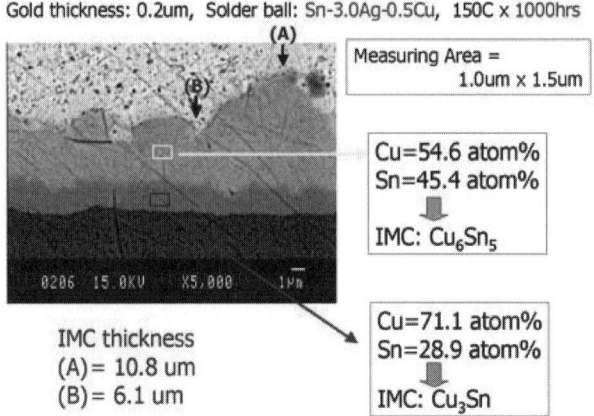

Gold thickness: 0.2um, Solder ball: Sn-3.0Ag-0.5Cu, 150C x 1000hrs

(A)

Measuring Area =
1.0um x 1.5um

(B)

Cu=54.6 atom%
Sn=45.4 atom%
IMC: Cu_6Sn_5

0206 15.0KV X5,000 1µm

IMC thickness
(A)= 10.8 um
(B)= 6.1 um

Cu=71.1 atom%
Sn=28.9 atom%
IMC: Cu_3Sn

Fig. 9.27. IMC formed between immersion gold surface finish and Sn-3.0Ag-0.5Cu Pb-free solder alloy after high-temperature aging at 150°C for 1000 hours showing IMC thickness range of approx. 6.1µm to 10.8 µm (Source: Uyemura)

Process Description

Immersion gold is a relatively simple process, as shown in Table 9.18.

Table 9.18. Immersion Gold – Typical Process Sequence

Process Step	Treatment Temperature (°C)	Treatment Time (sec)	Function
Clean	40-60	5 min	Remove copper oxide and solder mask residue
Rinse	RT	60 sec	
Microetch	20-30	2 min	Provides defined roughness by exposing the copper grain structure
Rinse	RT	60 sec	
Acid Dip	20-30	60 sec	Prepares copper surface for subsequent immersion step and protects from drag-in
Rinse	RT	60 sec	
Immersion Gold	80-90	15-30 min	Displacement of copper ions with gold
Rinse	RT	60 sec	
Dry			

Table 9.19. Direct Gold – Process Summary

Benefits	Concerns/Limitations
+ Simple process/suitable for horizontal application	– Relatively little production experience as a surface finish
+ Suitable for Pb-free soldering	– IMC formation between gold and tin ($AuSn_4$)
+ Good planarity	– Copper migration through the thin gold
+ Relatively low cost	– Copper re-deposition from immersion gold bath
	– Relatively short shelf life (<5 months) due to porosity of thin gold
	– Process chemistry cost (precious metal)

Unlike the other copper/tin IMC processes (Pb-free HASL, OSP, immersion silver and immersion tin), the immersion gold finish offers the possibility to deposit additional gold to make the surface suitable for gold wire bonding.

Key Issues - Immersion Gold

Direct gold (also known as direct immersion gold) is a relatively new process with very limited production experience. The process would appear to offer a simple method for providing a solderable finish that meets Pb-free requirements. However, there are shelf life constraints due to the extremely thin nature of the deposit. Copper migration can occur during storage which can negatively impact solderability. Also, the question of $AuSn_4$ IMC formation should be considered. Although the thin nature of the deposit would seem to limit the amount of such IMC, the issue remains regarding the possible re-deposition at the solder/pad interface. Generally, direct gold is another example of a surface finish that is most adversely affected by the relative lack of experience and use within the industry. Table 9.19 provides a summary of the primary benefits and concerns regarding the direct gold finish.

9.2.10 Custom Finishes

For certain technology applications, there is a need for more than one surface finish. The PWBs for some mobile phones are an example, combining both ENIG and OSP. For suitable resistance to wear and corrosion, the contact switch pads are plated with ENIG as a durable finish with excellent conductivity properties. Other pads requiring surface mount and micro-BGA components need a reliable solderable joint to meet drop-test standards.

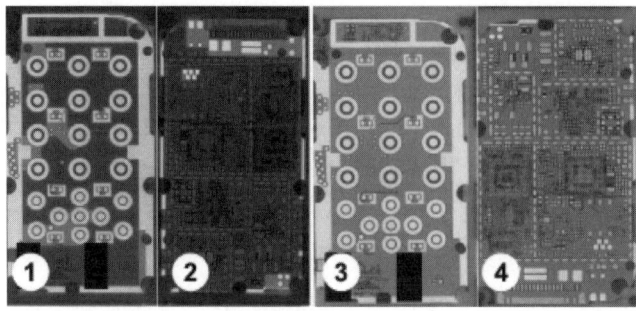

Fig. 9.28. Example of mobile phone PWB with ENIG on contact switch pads and photoresist covering features for OSP application, 1 and 2, and photoresist stripped showing OSP treated pads, 3 and 4 (Source: Atotech)

The use of OSP on these pads produces a copper-tin IMC, resulting in a less brittle solder joint.

After solder mask application, a secondary imaging technology (SIT) is typically required, whereby a photoresist is first imaged (only on those specific areas where OSP will be applied) and ENIG is then deposited, primarily on the contact switch pads. For such applications, a special photoresist is required because it must withstand the extended high-temperature processing of the ENIG process. Following ENIG, the photoresist is removed and OSP is then deposited on the exposed copper. The selective nature of the OSP will not allow it to deposit on the ENIG finish. Figure 9.28 depicts the typical progression of the surface finish application for mobile phone PWBs. An overview of the benefits and concerns of the ENIG/OSP custom finish is summarized in Table 9.20.

Table 9.20. ENIG/OSP Custom Finish – Process Summary

Benefits	Concerns/Limitations
+ Low-cost and good Pb-free solderability	– Same concerns noted for OSP
+ Minimizes use of higher cost finish (ENIG)	– Same concerns noted for ENIG
+ Good planarity	– Relatively complex process (requires secondary imaging step)
+ Shelf life of 6-12 months (with proper storage)	– Discoloration of gold finish by OSP process
+ Combines benefits of wear resistance with solder joint performance	– Galvanic etching of copper surface during OSP processing

Most of the benefits and concerns that were previously described for both OSP and ENIG are equally valid for this custom application. In addition, sequentially combining ENIG and OSP in the PWB fabrication process can offer some challenges. During the copper microetching (pretreatment) for OSP deposition, a galvanic reaction can occur on copper pads electrically connected to ENIG-plated pads, leading to excessive etching of those copper pads. Also, use of conventional OSPs can cause discoloration of the immersion gold surface; requiring new OSPs to be developed.

Similar finish combinations can also be used for other applications in which OSP is applied to pad areas for surface mount components, while electrolytic nickel/electrolytic gold is deposited in the through-holes, adding a barrier layer to protect against dissolution of the copper, particularly important in Pb-free rework situations.

Summary

Several PWB surface finishes are commercially available to meet the requirements of Pb-free electronics manufacturing. In this chapter, a description of the most technically feasible alternatives has been presented. Benefits and concerns that are specific to each finish (or group of finishes) have been discussed in those respective sections. It must be noted that no surface finish has yet been developed that meets the requirements of Pb-free assembly while eliminating all of the well known processing, storage and operational concerns. As explained herein, to some extent each finish is adversely affected by one or more technological or cost issues. In addition, there are some processing, performance or reliability problems that affect all or many of the available finishes. For example, it is well known that Pb-free solders are more aggressive to copper in comparison to eutectic SnPb solder. Pb-free soldering on Pb-free HASL, OSP, immersion silver, immersion tin and immersion gold results in the formation of a Cu-Sn IMC, which reduces the underlying pad and though-hole copper thickness. With no barrier layer, any further thermal exposure (i.e. additional soldering operations or high-temperature application) will continue to dissolve copper into the solder joint forming more IMC. Such dissolution can be of particular concern where the initial copper is thin, such as controlled impedance applications. Wave soldering can be very aggressive, particularly during rework soldering, causing drastic reductions in copper thickness, most notably at the through-hole entrance or "knee."

Voiding is another example of an issue that affects soldering on all surface finishes to some extent. Previously in this chapter, the specific issue of microvoiding with respect to immersion silver was examined in some detail. However, the occurrence of voiding within the solder joint is not limited to this particular surface finish. Figure 9.29 shows the results of one investigation of the occurrence of "process voids" as a function of surface finish. As indicated, although immersion tin and Pb-free HASL exhibit a lower incidence of process voiding, the issue does affect soldering on each of the surface finishes to some extent [36].

The action of voiding is mainly attributable to the flux outgassing within the solder joints when the solder is in the molten state, caused by vaporization or thermal decomposition of flux ingredients and/or reaction products formed due to fluxing. Certain conditions lead to voiding that have little to do with the board surface finish, including the thickness and activity level of the flux, with increases in voiding occurring with less active, thinly deposited fluxes. The void content also increases with increasing metal content of solder paste, partly due to an increase in solder powder oxide, partly due to an increasing difficulty for the flux to escape due to tighter powder packing and the formation of a greater amount of high viscosity metal salt. Solderability issues related to the board or component surface finish can also consume the flux in the solder paste or during wave soldering leading to process voiding.

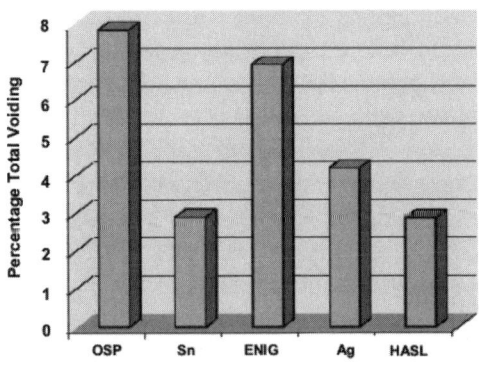

Fig. 9.29. Average percentages of total voiding of soldered BGA component solder balls by board finish type (Source: Dage)

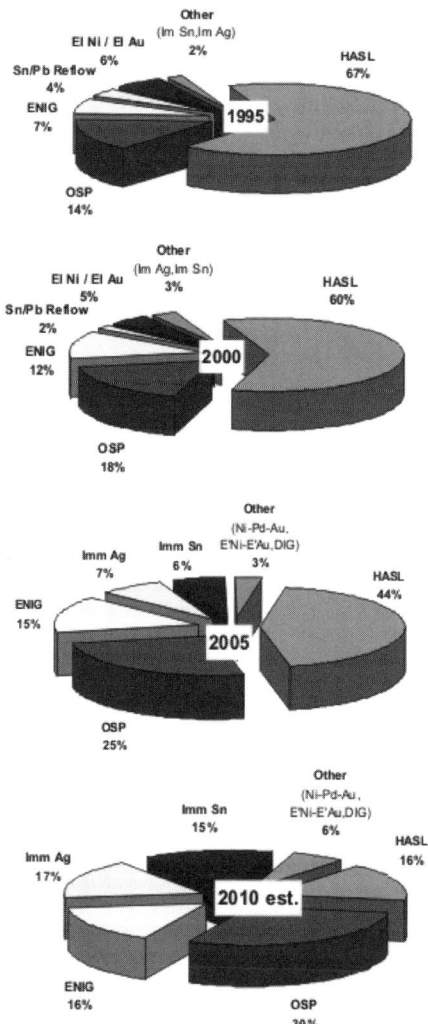

Fig. 9.30. Estimated distribution of major types of surface finishes for PWB applications, 1995-2010 (Source: Atotech)

Perhaps no other segment of PWB fabrication has been impacted more by the requirements of RoHS and similar Pb-free initiatives than that of board surface finishes. Until recently, eutectic Sn/Pb HASL was the surface finish of choice for the majority of applications. Coupled with the need for thinner and more planar coatings, the Pb-free requirements have further impacted the distribution of PWB surface finishes used in the industry. Figure 9.30 shows a graphical interpretation of the evolution of the PWB surface finish market from 1995 through 2010 (estimated). The shift from eutectic Sn/Pb HASL to the alternative finishes shown in the figure is a result of these changing environmental and technical requirements.

Tables 9.21 and 9.22 represent an attempt to summarize the most important surface finish evaluation criteria. To simplify the evaluation, the tables represent Pb-free issues pertaining to both PWB fabrication (Table 9.21) and assembly and manufacturing (Table 9.22). It is important to note that this summary is a subjective analysis based on the information compiled and included in this chapter. This information is not intended as a specific guide for selecting any surface finish to meet a particular design requirement or assigned function. The examination is only meant to serve as an overall assessment of the various surface finishes and their respective value or performance capability within each category.

Table 9.21. Summary of PWB Fabrication Issues for Pb-free Applications

	Pb-free HASL	OSP (high temp)	Imm. Ag	Imm. Sn	ENIG	Ni/Pd/Au	El Ni/ El Au	Eless Ni/ Eless Au	Direct Au
Additional thermal stress	— —	0	0	0	0	0	0	0	0
PWB thickness (>2.5mm)	— —	0	0	0	0	0	— —	0	0
PWB thickness (<1.0mm)	— —	0	0	0	0	0	0	0	0
Visual inspection	+	— —	+	+	+	+	+	+	−
Thickness variation	— —	−	0	+	+	+	−	+	+
Surface cleanliness	— —	0	−	−	0	0	0	0	0
Solder mask compatibility	— —	0	−	−	0	0	0	— —	0
Production experience	−	+ +	+ +	+ +	+ +	−	+ +	0	— —
Process operating window	+	+	+	+	−	−	+	— —	−
Handling and storage	++	−	−	−	+	+	++	++	— —

++ Major benefit + Some Benefit 0 No benefit or concern
− Some Concern − − Major Concern

Table 9.22. Summary of Assembly/Manufacturing Issues for Pb-free Applications

	Pb-free HASL	OSP (high temp)	Imm. Ag	Imm. Sn	ENIG	Ni/Pd/Au	El Ni/ El Au	Eless Ni/ Eless Au	Direct Au
Solderability - 1 reflow (air)	++	++	++	++	++	++	++	++	+
Solderability - 2 reflows (air)	+	+	+	+	++	++	++	++	–
Solderability 2 reflows + wave (air)	0	–	0	0	++	++	++	++	–
Shelf life (12 months)	0	0	0	0	++	++	++	++	– –
Fine-pitch/SMT Suitability	– –	++	++	++	++	++	0	++	++
IMC Mechanical Strength	0	0	0	0	–	–	–	–	–
Copper dissolution	– –	–	–	–	+	+	+	+	–
Wire bonding (Al)	no	no	yes*	yes*	yes	Yes	No	No	no
Wire bonding (Au)	no	no	No	no	no	Yes	Yes	Yes	no
Contact switching	no	no	yes**	no	yes	Yes	Yes	Yes	no
ICT Probe Suitability	++	– –	0	+	++	++	++	++	–

++ Major benefit + Some Benefit 0 No benefit or concern
– Some Concern – – Major Concern
* For thick Al-wire only (>240um) **Switch design dependent

It is important to note that IPC design standards are not changing in response to Pb-free requirements. Land patterns in the newly released IPC-7351A meet requirements for surface-mount assembly with either SnPb or Pb-free solders [37]. When IPC-7351 design rules are applied, Pb-free SMT assembly do not typically differ from eutectic SnPb surface mount soldering, and dimensional modifications are not required. In particular, for PCB surface finishes, fabricators have many options, and nearly all will work for Pb-free assemblies. The choice of finish is dependent on the application and preference. Pb-free HASL, OSP, immersion silver, immersion tin, and electroless nickel/immersion gold (ENIG) have all been used successfully in Pb-free assemblies.

IPC has published two board surface finish standards, IPC-4552 [22] for ENIG and IPC-4553 [16] for immersion silver. A third specification, IPC-4554 [16], addressing immersion tin is in its fourth draft version and nearing release. A fourth standard in the series, IPC-4555 will set forth standards for OSP and is currently underway. J-STD-003A, *Solderability Tests for Printed Boards,* is currently being updated to include testing for Pb-free board finishes [38].

The recently adopted requirements for Pb-free electronics will play a significant role in the selection of a PWB surface finish. As more features and functional requirements are built into the various electronic devices, continued miniaturization of electronics packaging and interconnection circuitry is a certainty. As the example in Figure 9.31 shows, as feature dimensions continue to decrease, performance and functional reliability become even greater concerns. To keep pace with the technology requirements, the more promising surface finish processes will be those that operate with limited interference from physical, chemical and mechanical limitations. Selection of the proper PWB surface finish will become more critical with these increasing demands, particularly in light of the never-ending drive to reduce costs. As a result, each finish must be examined and evaluated based on its recognized functional capabilities and known potential weaknesses.

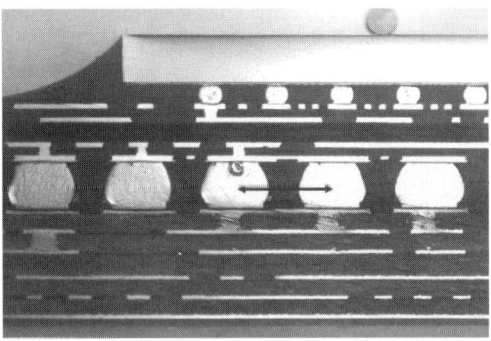

Fig. 9.31. Example of 0.5mm solder ball pitch in graphics chip assembly of FC-CSP for Fujitsu F900 mobile phone (Source: Prismark/Binghamton University)

Acknowledgements

The authors would like to acknowledge the contributions of members of the Surface Finish Business Technology Team of Atotech Deutschland GmbH and Jasbir Bath of Solectron, Milpitas, CA for their assistance and support in the preparation of this chapter.

References

1. Cemco-FSL (2005) The Newest Surface Finish Alternative: LEAD-FREE HASL - Its Development and Advantages.
2. Sweatman, K (2006) Fact and Fiction in Lead-free Soldering. Global SMT & Packaging.
3. Sweatman, K, Nishimura, T (2006) The Fluidity of the Ni-Modified Sn-Cu Eutectic Lead Free Solder. Proceedings of IPC Printed Circuits Expo.
4. Fellman, J and Lee, S (2005) FAQs about Lead-free HASL; Circuit Board Technologies (www.electronicmaterials.rohmhaas.com).
5. Vianco, P (1998) An Overview of Surface Finishes and Their Role in Printed Circuit Board Solderability and Solder Joint Performance. Circuit World.
6. Sikorcin, G (2005) Lead-free HASL. Florida CirTech PowerPoint presentation. SMTA Atlanta Chapter.
7. Sun, S., et al. (2006) Novel OSP Coatings for Lead-free Processes. Circuitree.
8. Geiger, D and Shangguan, D (2005) Evaluation of OSP Surface Finishes for Lead-free Soldering. Proceedings of SMTA International.
9. O'Connell, J (2004) Study and recommendations into using lead free printed circuit board finishes at manufacturing in circuit test stage. (www.agilent.com)
10. Lopez, Edwin et al. (2005) Effects of Storage Environments on the Solderability of Immersion Silver Board Finishes with Pb-based and Pb-free Solders. SMTA Journal, Vol. 18, Issue 4.
11. Cullen, D (2004) The Effects of Immersion Silver Circuit Board Surface Finish on Electrochemical Migration. Proceedings of SMTA International.
12. IPC-4553 (June 2005) Specification for Immersion Silver Plating for Printed Circuit Boards.
13. Veale, Robert (2005) Reliability of PCB Alternate Surface Finishes in a Harsh Industrial Environment. Proceedings of SMTA International.
14. Cookson Electronics (2005) AlphaSTAR – Enthone's Newest Generation Immersion Silver for PWB.(www.galvanord.dk)
15. IPC-4554 (March 2005) Specification for Immersion Tin Plating for Printed Circuit Boards, 4th Draft.
16. JEDEC/IPC JP002 (March 2006); Current Tin Whiskers Theory and Mitigation Practices Guideline.

17. Merix Corporation (2006) Comparison of Electroless Ni/Immersion Au vs. Electrolytic Ni/Au. (www.merix.com).
18. Johal, K, et al. (2004) Impacts of Bulk Phosphorous Content of Electroless Nickel Layers to Solder Joint Integrity and their Use as Gold- and Aluminum-Wire Bond Surfaces. Proceedings of SMTA Pan Pacific Microelectronics Symposium.
19. Fields, RJ, Low, SR and Lucey, GK (Oct 1991) Physical and Mechanical Properties of Intermetallic Compounds Commonly Found in Solder Joints. *The Metal Science of Joining*, Cincinnati, OH, pp. 165-173.
20. Toleno, B (2005) PCB Surface Finish Options for Lead-free Manufacturing. EMSNow.
21. Jacobson, DM and Humpston, G (1989) Gold Coatings for Fluxless Soldering. Gold Bull.
22. Mei, Z, Kaufmann, M, Eslambolchi, A and Johnson, P "Brittle interfacial fracture of PBGA packages soldered on electroless nickel/immersion gold", Proceedings 48 ECTC, May 1998, p. 952.
23. Johal, K (2001) Study of the Mechanism Responsible for Black Pad Defect on Assembled Electroless Nickel/Immersion Gold PWBs. Proceedings of IPC Printed Circuits Expo.
24. Chan, K, Kwok, WM, Bayes, M (2004) A new generation of electroless nickel immersion gold (ENIG) process for PWB surface finishes, KWPCA Journal, No. 13, pp. 1-7.
25. IPC-4552 (Oct 2002) Specification for Electroless Nickel/Immersion Gold (ENIG) Plating for Printed Circuit Boards, Appendix 5.
26. Oda, Y, Masayuki, K, Hashimoto, S (2006) IMC Growth Study on Ni-P/Pd/Au Film and Ni-P/Au Film Using Sn/Ag/Cu Lead-free Solder. Proceedings of IPC Printed Circuits Expo.
27. Lamprecht, S et al. (2005) Implementing Green PCB Production Processes. The Board Authority, Vol. 6, No. 1.
28. Milad, G and Gudeczauskas, D (2006) Surface Finishes for RoHS Compliance. Board Challenges (and Solutions) for Next Generation Designs; PBR Seminars.
29. Tsukada, K (2005) Development of New Surface Finishing Technology for PKG Substrate with High Bondability. Proceedings of IEEE/CPMT Advanced Packaging Materials Symposium.
30. Johal, K, Roberts, H and Lamprecht, S (2004) Electroless Nickel/Electroless Palladium/Immersion Gold Process For Multi-Purpose Assembly Technology. Proceedings of SMTA International.
31. Coombs, CF Jr. (1996) Printed Circuit Handbook, Fourth Edition; McGraw Hill Publishing.
32. Denman, RD (1996) Soldering to Gold Coatings, ITRI Publication, No. 736.
33. Hashimoto, S, Kiso, M, Oda, Y, Otake, H, Milad, G, and Gudaczauskas, D; (2006) Direct immersion gold (DIG) as a final finish, Proceedings of IPC Printed Circuits Expo.

34. Milad, G and Gudeczauskas (2006) Solder Joint Reliability of Gold Surface Finishes (ENIG, ENEPIG and DIG) for PWB Assembled with Lead Free SAC Alloy. Uyemura Library (www.uyemura.com).
35. Walsh, D, Milad, G and Gudeczauskas, D (August 2004) Printed Circuit Boards: Final Finish Options. PF Online. (www.pfonline.com))
36. Bernard, D and Bryant, K (2004) Does PCB Pad Finish Affect Voiding Levels in Lead-Free Assemblies? Proceedings of SMTA International.
37. IPC-7351, Generic Requirements for Surface Mount Design and Land Pattern Standard.
38. IPC Designers Council (September 2005); Lead Free: How Will IPC Standards Change?

Chapter 10: Lead-Free Soldering Standards

Jasbir Bath, Solectron Corporation, Milpitas, California

10.1 Introduction

This chapter will review the current and developing standards, which affect lead-free soldering. Developments in standards for lead-free soldering are not complete and are progressing with new discoveries and developments. The majority of the chapter will cover IPC standards with reference to other standards, which are considered relevant and known.

This chapter is broken down into four sections that are:

1. IPC and JEDEC Assembly Standards
2. IEC Standards
3. Japan Standards
4. Other Standards

10.2 IPC and JEDEC Standards

IPC and JEDEC standards are well known and well adopted standards across the global electronics industry. The following section will explore the development of certain IPC and JEDEC standards with respect to lead-free soldering.

10.2.1 Assembly Standards

There are various assembly standards used in the industry with the best known being J-STD-001D.

10.2.1.1 J-STD-001D

The J-STD-001D standard [1] includes criteria for materials, methods and verification for producing quality soldered interconnections and assemblies. It includes among other areas visual inspection and SIR test requirements for lead-free and tin-lead assemblies and indicates that the primary difference between lead-free solder alloys and traditional tin-lead alloys is visual appearance of solder. The lead-free solder can appear duller or have a more grainy appearance than a tin-lead joint which is due to surface roughness of the high tin lead-free solder joint. A duller joint is not grounds for rejection of a lead-free or a tin-lead joint. Lead-free solder joint pictures have been inserted side by side with tin-lead solder joint pictures in the standard for comparison.

10.2.2 Acceptance Standards

10.2.2.1 IPC-A-610-D

The IPC-A-610D standard [2] similar to the J-STD- 001D process standard indicates that the primary difference between lead-free solder alloys and traditional tin-lead alloys is visual appearance of solder. The IPC-A-610D standard indicates that the preferred solder connections require low to near zero contact angle with the target solder connections having a shiny to satin luster. Inspection of lead-free solder connections may show improper rejections of acceptable lead-free joints with texture and contact angles different from tin-lead.

This standard also gives criteria for fillet lifting for lead-free and accept/reject criteria for shrinkage grooves for lead-free as well as criteria for voiding for tin-lead and lead–free BGA/CSP joints. With lead-free BGAs/CSPs, there may be increased voiding compared with tin-lead but the 25% void area criteria would still be used. Other topics include component orientation and soldering criteria for through-hole, SMT and discrete wiring assemblies, mechanical assembly, cleaning, marking, coating, and laminate requirements.

10.2.3 Solderability Standards

Solderability standards are used to assess the solderability of bare components and bare boards with J-STD-002B [3] being the most used standard for the component assessment and J-STD-003A [4] for bare boards.

10.2.3.1 J-STD-002B

IPC/EIA/JEDEC J-STD-002B standard provides a means to assess solderability of electronic component leads, terminations, solid and stranded wire. The standard includes test method choices, defect definitions, acceptance criteria and illustrations. It evaluates solderability testing for component leads and terminations and looks at wetting balance and dip and look tests for evaluations on solderability. It gives the maximum impurity levels for the tin-lead pot and specifies the type of flux to use (ROL1). It also gives the pre-conditioning steam ageing conditions before solderability testing on components.

The new revision of J-STD-002 (Revision C) which is in final draft form indicates the use of Sn3Ag0.5Cu solder for lead-free component testing and gives impurity levels for Sn3Ag0.5Cu for the pot. The flux type to use during wetting balance and dip and look testing is ROL1 (the same as with tin-lead solder).

10.2.3.2 J-STD-003A

IPC/EIA J-STD-003A provides industry-recommended test methods, defect definitions and illustrations to assess the solderability of printed board surface conductors, lands and plated-through holes. The test methods covered include edge dip, rotary dip, solder float, wave solder and wetting balance. Only tin-lead solder is used at the present time with ROLO type flux used. The impurity level of the solder used during testing are the same as J-STD-002.

10.2.4 Assembly Materials

Assembly standards are used to ensure the quality and impurity levels of incoming solder materials. For incoming solder materials the J-STD-006B standard is used [5].

10.2.4.1 J-STD-006B

This standard indicates the nomenclature, requirements and test methods for electronic grade solder alloys, for fluxed and non-fluxed bar, ribbon, and powder solders. This standard contains the maximum impurity levels for a variety of lead-free solders including SnAgCu. The most significant change from the previous revision was to indicate that the maximum lead impurity for a lead-free solder alloy is 0.1wt% from the previous standard version of 0.1 to 0.2wt% lead as the 0.1wt% lead impurity level in lead-free solder would be in line with the European Union ROHS legislation.

10.2.5 Assembly Support Standards

Assembly support standards are supplementary standards such as IPC7711/7721 [6] and IPC1066 [7] and JESD-97 [8], which are discussed, in further detail in the following sections.

10.2.5.1 IPC 7711A/7721A Rework, Repair and Modification of Electronic Assemblies

This standard at the present time is still concentrating on tin-lead rework. More emphasis in the future will need to concentrate on lead-free rework in particular to lead-free mini-pot rework and BGA/CSP rework which are discussed in more detail in the rework chapter of this book.

10.2.5.2 IPC 1066 and JEDEC Standard JESD97 on Lead-Free Labeling

IPC 1066 [7] establishes the requirements for distinctive symbols and labels to be used to identify materials that are lead free (Pb-free) and capable of providing lead-free second level interconnects, and for indicating certain types of lead-free materials and the maximum allowable soldering temperature. It also establishes the requirements for labeling a bare board if the base resin is halogen free and the type of conformal coating used after assembly. Both IPC 1066 and JEDEC JESD-97 [8] standards indicate lead-free labeling of components for 2nd level termination finishes and solders used on boards for lead-free 2nd level assembly.

Identification of the lead-free materials used is critical for those performing assembly, rework, repair and recycling operations. These standards make available a method for marking and labeling of lead-free components and assemblies. The most frequently used lead-free solder alloy for assembly and lead-free BGA/CSP components is SnAgCu, which is designated by e1 on the board or component. The most commonly used lead-free termination finish is pure Sn, which is designated by e3 on the component body. A listing of the e codes are below. The 'e' code e2 indicates SnAg or SnCu alloys but not alloys of SnAgCu.

e1 – SnAgCu (shall not be included in category e2)
e2 – Sn alloys with no Bi or Zn excluding SnAgCu
e3 – Sn
e4 – Precious metal (e.g. Ag, Au, NiPd, NiPdAu) (no Sn)
e5 – SnZn, SnZnX (all other Tin Zinc or Tin Zinc and other tertiary alloys not containing Bi)

e6 – contains Bi

e7 – low temperature solder ($\leq 150^\circ C$) containing Indium (no Bi)

For example an e1 code on a BGA component would indicate the BGA sphere composition was SnAgCu where an e1 code on the board would indicate that the solder material to use for assembly would be SnAgCu. The new J-STD-609 standard [9] which at the time of writing of this chapter is in draft form is meant to combine the existing IPC 1066 and JEDEC JESD-97 standards whilst also including identification of board surface finishes and board laminate types as well as tin-lead components and solders.

10.2.5.3 IPC-7095 (BGA Standard)

IPC 7095 Revision A [10] gives information to those currently using or considering a conversion to area array packaging. It provides guidelines for BGA inspection and repair and also addresses reliability issues. It contains some information on lead-free soldering including a typical lead-free reflow profile. The new revision of this document which is currently being drafted will contain an increased amount of information on lead-free. Currently lead-free water soluble soldering pastes show increased voiding which is a concern especially if current BGA/CSP voiding criteria was adhered to. This increased voiding may be due to the development nature of these pastes, which are typically used in high reliability applications, which are typically exempt from the current European Union ROHS legislation.

10.2.5.4 IPC-9701A (Reliability Standard)

This standard [11] provides specific test methods to evaluate the performance and reliability of surface mount solder attachments of electronic assemblies. The current Revision A of this standard discusses lead-free reliability testing in more detail. It still recommends the typical 10 minute dwell times at the temperature extremes for the most common $0^\circ C$ to $100^\circ C$ ATC testing which is similar to tin-lead reliability testing but does give the option for increased dwell time testing for lead-free solders. It also indicates that there are undeveloped lead-free acceleration models at this time.

10.2.6 Component Standards

Component standards were some of the first to be updated for lead-free in certain areas because of the need to have higher temperature rated components for lead-free assembly. With increasing production data, these standards are being further updated.

10.2.6.1 J-STD-020C (Component Temperature Rating)

This standard [12] identifies the classification level of nonhermetic solid state surface mount devices that are sensitive to moisture-induced stress. It is used to determine what classification level should be used for initial reliability qualification. These devices can be properly packaged, stored, and handled to avoid subsequent thermal/mechanical damage during solder reflow attachment and/or repair operations.

Developed by IPC and JEDEC, the J-STD-020C standard has temperature ratings for moisture sensitive components for lead-free soldering These range from 245°C to 260°C dependent on component thickness and body size. Even if the lead-free rated component is not rated to 260°C, due to its component thickness or body size, it has to still be able to simulate 1X profile at 260°C peak to be compatible for lead-free area array component rework. The new draft revision of this standard is looking to develop temperature rating criteria for other component devices such as passive components.

10.2.6.2 J-STD-033B (Component Bake Out Requirements)

Developed by IPC and JEDEC this standard [13] gives Surface Mount Device manufacturers and users standardized methods for handling, packing, shipping and use of moisture/reflow sensitive SMDs. These methods help avoid damage from moisture absorption and exposure to solder reflow temperatures that can result in yield and reliability degradation. The baking temperatures and times associated with the specific moisture sensitivity level are the same for both lead-free and tin-lead temperature rated components.

10.2.6.3 JEDEC JESD22-B106C and JEDEC JESDD22-A111 (Component Resistance to Wave Soldering Temperatures)

The JEDEC JESD22-B106C method [14] establishes a standard procedure for determining whether through-hole solid state devices can withstand the effects of the temperature to which they will be subject during soldering of their leads. This document only covers through hole wave soldered components. It does not address components glued to the backside (bottom side) of the board and soldered using the wave soldering process.

The current revision of this standard indicates a 260°C temperature rating for lead-free wave soldering through-hole connectors for 10 seconds (dual wave: chip and main wave), which is the same as that indicated for tin-lead wave soldering. Considering that for wave and rework wave soldering the lead-free wave soldering temperatures are reaching 270°C, the rating for lead-free wave soldered through-hole components needs adjustment, which is in progress. This document has also not previously included the implications of the preheat temperature prior to wave soldering which for lead-free wave soldering may increase in temperature compared to tin-lead.

The JESD22-A111 standard [15], which covers bottom-sided wave soldered components, has yet to be revised for lead-free wave soldering temperatures. The current standard indicates only tin-lead wave soldering temperatures at 260°C for 10 seconds (dual wave: chip and main wave) which will be revised for lead-free. The purpose of the JESD22-A111 test method is to identify the potential wave solder classification level of small plastic Surface Mount Devices (SMDs) that are sensitive to moisture-induced stress so that they can be properly packaged, stored, and handled to avoid subsequent mechanical damage during the assembly wave solder attachment and/or repair operations. This test method also provides a reliability preconditioning sequence for small SMDs that are wave soldered using full body immersion.

10.2.6.4 JEDEC Tin Whisker Methods and Acceptance Criteria

Tin whisker standards have been developed as a direct result of concern over the increased supply and use of pure tin and lead-free high tin containing component terminations. There are 2 main testing standards, which are used with an additional guideline publication discussing tin whisker mitigation techniques.

JEDEC Standard JESD 22A121

This standard [16] contains three types of tin whisker test conditions to use as well as indicating component inspection number and inspection method. It also contains typical component preconditioning tin-lead and lead-free reflow profiles prior to tin whisker testing.

JEDEC JESD-201 Standard

This JEDEC JESD-201 tin whisker test method standard [17] acts as a compendium to the JEDEC 22A121 standard in that it indicates and in some cases updates the three tin whisker test conditions and includes acceptance tin whisker length criteria. It also suggests test method times and criteria to use based on whether a technology or a manufacturing acceptance qualification is required.

JP-002

This document [18] provides some insight into the theory behind tin whisker formation as it is known today and, based on this knowledge, potential mitigation practices that may delay the onset of, or prevent tin whisker formation. The potential effectiveness of various mitigation practices are briefly discussed. References behind each of the theories and mitigation practices are also provided.

10.2.7 Rigid Laminate Standards

Board laminate standards are probably one of the least developed standards for lead-free as the initial push to lead-free soldering was with lead-free cell phones and laptop consumer boards, which are small and thin types of boards which typically did not reach excessive lead-free soldering temperatures. With the migration to lead-free soldering for thicker boards such as server type assemblies there is a need for higher temperature rated laminate boards for lead-free assembly. With increasing prototype and production data, these standards are being updated.

10.2.7.1 IPC-4101B

This specification [19] covers the requirements for base materials (laminate and prepreg) to be used primarily for rigid or multilayer printed boards for electrical and electronic circuits. This standard provides additional information and data on printed circuit board materials that are better able to withstand the newer assembly operations employing higher thermal exposures, including those assembly practices that utilize lead free solders.

The standard contains six specification sheets for FR4 lead-free temperature rated materials. The specification sheets give information on decomposition temperature, Z-axis expansion, T260, T288, T300 and CAF resistance. These materials are also required to be UL-94-V0 flammability rated. Lead-free specifications sheets require improved thermal resistance, Z-axis expansion and inter-laminar adhesion. The new draft revision of this standard is looking to add specification sheets for FR4 lead-free temperature rated materials which are low halogen containing.

10.2.7.2 IPC-7351

The document [20] covers land pattern design for passive and active components. It includes updates to existing sections addressing land pattern guidelines for wave or reflow soldering, via location guidelines, fiducials and courtyard boundaries.

The new draft of IPC-7351 makes some references to lead-free in terms of temperature rating testing temperatures and does not indicate any changes in surface mount land pattern designs for lead-free compared with tin-lead. There will need to be more input and work into the development of complete design guidelines for lead-free soldering including wave and rework considerations.

10.2.7.3 IPC Board Surface Finish Standards

In response to a need to understand and determine the best choice to make regarding lead-free board surface finish there are a number of new or developing standards in this area. IPC 4552 standard [21] is a specification for ENIG (NiAu) board finish and IPC 4553 standard [22] is a specification for Immersion Silver board finish. The IPC-4553 specification indicates metrics for immersion silver deposit thickness based on performance criteria. Developing lead-free surface finish standards include a draft standard for immersion tin [23] and a draft standard for OSP [24]. Board surface finishes are discussed in more detail in the board surface finish chapter of this book.

10.2.8 Materials Declaration Standards (IPC-1065 and IPC 1752)

The IPC-1065 standard [25] was developed as a direct result of a need for consistent reporting data regarding the materials declaration of materials including European Union ROHS compliancy verification needs. The follow on IPC 1752 Materials Declaration Management standard [26] established the requirements for exchanging material and substance data between suppliers and their customers for Electrical and Electronic Equipment (EEE). Included with the standard is a User's Guide (IPC-1752-3) that describes how to use the two PDF reporting forms identified as IPC-1752-1 and IPC-1752-2. These forms can record in a standard format, the material declaration requirements for components, boards, sub-assemblies, and final products.

10.3 IEC Standards

IEC standards are basically global standards, which are developed from a national standard such as an IPC or a JEDEC standard. Probably the most interesting IEC standard that is being developed based on input from multiple countries is a standard for testing of substances in components and electronics assemblies to ensure European Union ROHS compliancy.

10.3.1 IEC 62321 Draft Standard: Evaluation of Materials for ROHS Compliancy Testing

This draft IEC 62321 standard [27] is developing test methods to analyze for lead, cadmium, mercury, hexavalent chromium, PBB (Polybrominated bi-phenyls) and PBDE (polybrominated di-phenlyethers). The test methods to determine these prohibited materials down to 0.01wt% for Cadmium or 0.1wt% for the other substances are still in the process of being developed. The materials under test in components are usually inhomogeneous which makes testing even more difficult. Currently round robin testing being conducted to validate the test methods employed appeared to indicate that the tests for hexavalent chromium, PBB and PBDE are not as reliable or repeatable as the tests for lead, cadmium and mercury. One of the options being discussed is to release a standard for lead, cadmium and mercury but release a guidance document for hexavalent chromium, PBB and PBDE until the issues with the test methods for these three substances are resolved.

10.4 Japan (JEITA) Standards

There are many standards in Japan, which in many respects tend to be similar to IPC and JEDEC standards. There are continuing efforts underway to ensure that there are consistent standards developed between IPC, JEDEC and Japan (JEITA). The following two standards are some of the most important.

10.4.1 JEITA Lead-Free Component Temperature Rating Standards

This standard [28] is very close to the IPC/JEDEC J-STD-020C standard for component temperature rating standard. There was an issue with certain component body and thickness temperature ratings, which were not previously consistent with J-STD-020C, but due to the co-operation of IPC, JEDEC and JEITA, consensus standards are being developed so that component suppliers from different countries do not test to different temperature ratings which would create confusion and increased cost.

10.4.2 JEITA Lead-Free Labeling Standards

This standard [29] has not been adopted by IPC and JEDEC with labeling in the JEITA guidance document using different labeling codes for lead-free solders (not 'e' codes as for IPC and JEDEC standards) with no labeling codes for components. There is also a different JEITA standard [30] to identify the presence of ROHS substances. There has been efforts to pursue a single global lead-free labeling standard but it seems the best compromise will be to have Japan companies using one lead-free labeling standard and North America and European companies using a different lead-free labeling standard. As long as there is some type of labeling for a lead-free soldered board and assemblers and repair personnel are aware of the different labeling types, then this should not cause too much disruption.

10.5 Other Standards

There are many standards from different parts of the world. The development of a standard from China is discussed which has implications based on the globalization of electronics product use.

10.5.1 China Draft Standard: EPUP (Environmental Protection Use Period)

This draft standard [31] discusses a requirement to assess and indicate the length of time before the ROHS substances (lead, cadmium, mercury, hexavalent chromium, PBB (Polybrominated bi-phenyls) and PBDE (polybrominated di-phenlyethers)) present in the electronic product are likely to start breaking down and leaching into the environment. Although the aim of the standard is good, it creates an unnecessary amount of labeling and interpretation for an electronics product, which will likely increase cost and confusion in the electronics industry.

Conclusions

This chapter reviewed some of the most common standards, which have been affected by the transition to lead-free soldering. Critical issues such as temperatures for lead-free soldering and visual inspection of lead-free soldered joints have been addressed to a certain extent but there are many more standards that need updating and this is being done with more data becoming available. There are additional standards being developed which are based nationally rather than globally which in some cases may complicate the transition to lead-free soldering.

Future Work

There are many areas for standards development for lead-free. There needs to be consideration and development of standards for multi-layer PCBs as current IPC standards only consider testing of the board laminate but not the final laminate assembled with copper layers which is representative of the product board itself. There needs to be development of more process related temperature standards. Component suppliers are not aware of the typical processing temperatures and times used for surface mount and wave assembly and for rework operations and understanding and determining these for lead-free soldering will help to develop component and board materials more resistant to the demands of higher temperature lead-free processing.

There should also be more consideration of avoiding duplication of effort with multiple standards being developed by multiple standards bodies around the world that leads to multiple tests being conducted by component or board suppliers that in many cases are not necessary and increase cost and confusion.

References

1. J-STD- 001D Standard (2005) Requirements for Soldered Electrical and Electronic Assemblies.
2. IPC-A-610D Standard (2005) Acceptability of Electronic Assemblies.
3. J-STD-002B Standard (2003) Solderability Tests for Component Leads, Terminations, Lugs, Terminals and Wires.
4. J-STD-003A Standard (2003) Solderability Tests for Printed Boards.
5. J-STD-006B Standard (2006) Requirements for Electronic Grade Solder Alloys and Fluxed and Non-Fluxed Solid Solders for Electronic Soldering Applications.
6. IPC-7711A/7721A Standard (2003) Rework, Repair and Modification of Electronic Assemblies.
7. IPC-1066 Standard (2004) Marking, Symbols and Labels for Identification of Lead-Free and Other Reportable Materials in Lead-Free Assemblies, Components and Devices.

8. JEDEC JESD-97 Standard (2004) Standard Marking, Symbols, and Labels for Identification of Lead (Pb) Free Assemblies, Components, and Devices.
9. J-STD-609 Draft Standard (2006) Marking, Symbols and Labels for Identification of Lead Free and Other Reportable Materials in Lead Free Assemblies, Components and Devices.
10. IPC-7095A Standard (2004) Design and Assembly Process Implementation for BGAs.
11. IPC-9701A Standard (2006) Standard Performance Test Methods and Qualification Requirements for Surface Mount Solder Attachments.
12. IPC/JEDEC J-STD-020C Standard (2004) Moisture/Reflow Sensitivity Classification for Nonhermetic Solid State Surface Mount Device.
13. IPC/JEDEC J-STD-033B Standard (2005) Handling, Packing, Shipping and Use of Moisture/Reflow Sensitive Surface Mount Devices.
14. JEDEC JESD22-B106C Standard (2005) Resistance to Soldering Temperature for Through-Hole Mounted Devices.
15. JEDEC JESD22-A111 Standard (2004) Evaluation Procedure for Determining Capability to Bottom Side Board Attach by Full Body Solder Immersion of Small Surface Mount Solid State Devices.
16. JEDEC JESD22A121 Standard (2005) Measuring Whisker Growth on Tin and Tin Alloy Surface Finishes.
17. JEDEC JESD-201 Standard (2006) Environmental Acceptance Criteria for Tin Whisker Susceptibility of Tin and Tin Alloy Surface Finishes.
18. IPC/JEDEC JP-002 Guideline (2006) Current Tin Whiskers Theory and Mitigation Practices Guideline.
19. IPC-4101B Standard (2006) Specification for Base Materials for Rigid and Multilayer Printed Board.
20. IPC-7351 Standard (2005) Generic Requirements for Surface Mount Land Pattern and Design.
21. IPC-4552 Guideline (2002) Specification for Electroless Nickel/Immersion Gold (ENIG) Plating for Printed Circuit Boards.
22. IPC-4553 Guideline (2005) Specification for Immersion Silver Plating for Printed Circuit Boards.
23. IPC-4554 Draft Guideline (2006) Draft Specification for Immersion Tin Plating for Printed Circuit Boards.
24. IPC-4555 Draft Guideline (2006) Draft Specification for OSP for Printed Circuit Boards.
25. IPC-1065 Standard (2005) Material Declaration Handbook.
26. IPC-1752 Standard (2006) Materials Declaration Management Standard.
27. IEC 62321 Draft Standard (2006) Draft Standard Procedures for the Determination of Levels for Six Regulated Substances (Lead, Mercury, Cadmium, Hexavalent Chromium, Polybrominated Biphenyls, Polybrominated Biphenyl Ether) in Electrotechnical Products, IEC TC 111 Working Group 3.
28. JEITA ED-4701 Method 301A Standard (2003) Resistance to Soldering Heat for Surface Mount Devices.

29. JEITA ETR-7021 Guidance Document (2004) Guidance for the lead-free marking of materials, components and mounted boards used in electronic and electrical equipment.
30. JIS C 0950 Standard (2005) The marking for presence of the specific chemical substances for electrical and electronic equipment.
31. Electronics Industry Draft Standard of the People's Republic of China (2006) General Rules for the Environmental Use Period of Electronic Information Products.

Conclusions

The legislation chapter reviewed the emerging wave of environmental regulations around the globe, which present technical, administrative, and procedural compliance challenges to producers. Failure to comply with these requirements also carries significant business, financial and legal implications. Companies must pay close attention to these evolving regulations and work with diverse experts to bring all the elements of sustained compliance together.

The alloy chapter reviewed various lead-free solder alloys with a focus on SnAgCu. It guided the reader to where future research on Pb-free alloys would be needed, which included:

- high reliability applications – server, military, aerospace, medical – for which exemptions are going away or where the lack of availability of tin-lead components will force a switch to lead-free soldering
- understanding whether there are any significant differences between different SAC alloy compositions in terms of reliability
- acceleration factors for SnAgCu and SnCu based alloys
- interactions of lead-free solder alloys and board/component surface finishes (containing gold, nickel, bismuth, silver, copper)
- effects of lead-free soldering time above liquidous temperature and peak soldering temperature on intermetallic formation and solder joint microstructure and its effect on reliability
- effect of lead-free cooling rate and subsequent aging on intermetallic compound and solder joint microstructure and their effect on reliability
- copper dissolution studies with different lead-free solder joint alloys and reliability of the resulting solder joint
- effect of CuSn, NiSn, NiCuSn, AgSn, AuSn intermetallic on lead-free solder joint reliability
- effect of lead-free surface mount, wave and rework (BGA/CSP, Wave, Hand) processing on intermetallic formation and solder joint microstructure with the resultant effect on reliability
- effect of lead-free wave solder holefill on reliability

- effect of ATC and mechanical testing (bend, shock, vibration) on lead-free solder joint reliability
- effect of ATC dwell time and temperature ranges on lead-free solder joint reliability

The SMT reflow chapter discussed the printability of lead-free SAC solder pastes which were shown to be equivalent to that of SnPb in most cases. With more production experience with lead-free assembly, solder paste suppliers would make more improvements in their flux formulations. There had been no real placement issues associated with lead-free components. Lead-free reflow soldering affected the visual appearance of soldered joints which was indicated in the IPC-610D standard. It also affected the temperature of the reflow profile which could cause issues for assembled components and boards due to the higher soldering temperatures. The surface mount component temperature rating standards had been adjusted to reflect these higher temperatures, but more work was needed for board temperature ratings for lead-free. X-ray images were fairly similar for lead-free and tin-lead soldered components.

The lead-free wave soldering chapter discussed the migration from the traditional tin-lead (Sn37Pb) alloy to lead-free solder with Sn3.0Ag0.5Cu as the main lead-free wave alloy alternative to align with the lead-free surface mount paste. Due to cost issues, alloys containing reduced or no silver, such as SN100C (Sn0.6Cu0.05Ni), could also be an alternative. The wave solder machine would need to be adapted to lead-free soldering to avoid erosion of all its machine parts that were in contact with the molten alloy which should be considered to be retrofitted.

Materials segregation to control lead-free solder from being mixed with tin-lead solder would be one of the most challenging items to implement moving forward. Frequency of bath alloy analysis would need to be reinforced with careful analysis of the elements such as lead, copper and iron. The lead-free wave process window would be narrowed and the risk of damage of components and/or PCB increased. Higher solids content no-clean fluxes would give good results for lead-free wave soldering but there would be an issue with test pin probeability results. If assemblers were not used to VOC-free fluxes, they would need education as they were typically preferred for lead-free soldering as they offered a complete green solution. With these water based fluxes, increased preheat would be more critical to ensure water removal prior to the wave soldering. Lead-free solder joints had a grainy appearance and some anomalies or defects may increase with lead-free soldering, including voiding, hot tearing and shrinkage grooves, fillet lifting, pad lifting and reduced hole fill. More studies would need to be performed on the final product application as the effect of these potential anomalies/defects on solder joint long term reliability was not well known.

The lead-free rework chapter discussed the choice of Sn3.5Ag for lead-free hand solder rework, although there had been movements to Sn3Ag0.5Cu to be the same alloy as used for surface mount and wave soldering. Typical soldering iron tip temperatures were found to be around 371°C (700°F), which was slightly higher than those used for tin-lead. The soldering tip would usually apply localized heating so that the component would be less likely to increase in temperature sufficiently to cause concern compared with lead-free BGA/CSP component rework. Good training would be needed to avoid excessive board and component damage. Soldering iron tip life would typically be degraded with lead-free high tin based solders with more care needed by the operators for solder iron tip maintenance.

For lead-free BGA/CSP Rework, temperatures being experienced were found to be higher than during 1st pass reflow. The peak temperature of the component during rework was found to be 15-25°C above that of the solder joint. A good temperature rating for components for lead-free soldering would be 260°C. The margin of error to maintain a lead-free minimum solder joint temperature of 230-235°C with a maximum body temperature of 245°C-250°C was very tight during BGA/CSP rework with the J-STD-020C standard helping to address this (260°C). In many cases, adjacent component temperatures were exceeded at 3.8mm (150mils) during rework on the board for tin-lead and lead-free BGA/CSP rework, which gave cause for concern for adjacent component temperatures and subsequent joint reliability. Rework equipment and nozzles were still in the process of development for lead-free with it generally being more difficult to control temperatures in rework than reflow with a need for more lead-free process manufacturing margin.

Bottom-side heat and thermal uniformity of BGA/CSP rework machines were critical to bring the board to proper lead-free rework temperatures. Rework machines needed more development with emphasis on optimized lead-free rework profiles, optimized rework machine development and machine repeatability. These were under investigation in the new iNEMI lead-free rework optimization project.

For lead-free hand PTH rework, there had been limited work done. The testing done so far indicated soldering temperatures and time may be slightly higher than for lead-free SMT hand soldering rework. In some cases, pre-heat of the board would be preferred. More work would be needed to develop the rework process with a general soldering iron tip temperature guideline of 371°C (700°F) to 427°C (800°F) for lead-free.

For lead-free PTH mini-pot rework, this area had minimal development. There were issues due to increased solder mini-pot temperatures and dwell times using lead-free high tin containing solders which lead to issues such as copper dissolution and reduced holefill. Alternative alloys such as Sn0.6Cu0.1Ni (SN100C) were being investigated to reduce the copper dissolution issue, but more work needed to be done investigating the type of copper plating used by the board supplier to understand if this could also reduce the amount of copper dissolution.

The lead-free reliability chapter indicated that along with SnPb assemblies, the assessment of lead-free solder joint reliability required detailed, quantitative investigations of the thermo-mechanical response of solder alloys under a wide range of stress, strain and temperature conditions. Lead-free solder joint reliability similar to SnPb reliability remained product- and application-specific. The case study indicated that lead-free solder joint life for a given product board relied on a large number of parameters describing thermal conditions, board and component geometry, and material properties. Significant progress had been made to quantify the thermo-mechanical behavior of SAC387/396 solders and board assemblies, including the development of test databases, life prediction models, and acceleration factors, and more work would be needed. Similar efforts would also be needed for each new solder alloy composition being considered for lead-free product assembly.

The backward compatibility and forward compatibility chapter reviewed the process and reliability implications of the mixing of tin-lead solder alloy and lead-free components and lead-free solder alloy with tin-lead components, with emphasis on the reliability of tin-lead solder and lead-free BGA/CSP backward compatible assemblies. There were conflicting experimental results on the reliability of BGA/CSP backward compatible assemblies. The majority of forward compatibility studies showed few or no issues, although excessive voiding of tin-lead BGA/CSP components with lead-free solder was a concern. The effect of lead content in mixed assemblies (forward compatibility and backward compatibility) was still questionable in terms of reliability. Data showed that the backward compatibility assemblies of chip components and lead-frame components were reliable in terms of solder joint integrity. The estimation of the liquidous temperature of mixed compositions in backward compatibility was presented for BGA/CSP, lead-frame and chip components. The estimation for BGA/CSP components could be used to guide the development of a reflow profile, but it needed to be noted that the estimation was an approximation and further experimental study would be needed to validate the accuracy of the method.

Both the backward and forward compatibility situations would need to be considered as transitional processes only with a full movement to lead-free

solder with lead-free components being the general goal to avoid any reliability issues associated with the two transition assembly situations. For lead-
free press-fit components, due to the press-fit material changing from SnPb
coated to lead-free coated, the process would need to be re-characterized.
Currently there were only limited studies done by OEM/EMS/suppliers.
No conclusion could be drawn yet in terms of the selection of the best PCB
surface finish, PCB laminate material, and compliant pin plating. It was
recommended that OEM and EMS providers and connector suppliers work
together to make the lead-free press-fit interconnection transition smooth,
without compromising quality and reliability.

The PCB laminate chapter discussed several laminate material properties
that were critical for evaluating laminate material choices for higher temperature lead-free assembly processing. Laminate material decomposition
temperature was the newest test method for evaluating thermal robustness,
and along with the overall Z-axis CTE test method was the most indicative
of what was required for both surviving higher temperature lead-free assembly processing and ensuring subsequent product reliability.

The PCB surface finishes chapter discussed the lead-free board surface
finishes which were commercially available to meet the requirements of Pb-
free electronics manufacturing. It gave a description of the most technically
feasible alternatives. Benefits and concerns specific to each finish (or group
of finishes) were discussed. It was noted that no surface finish had yet been
developed that met the requirements of Pb-free assembly while eliminating
all of the well known processing, storage and operational concerns. To
some extent each finish was adversely affected by one or more technological or cost issues. In addition, there were some processing, performance or
reliability problems that affected all or many of the available finishes. For
example, it was well known that Pb-free solders were more aggressive to
copper in comparison to eutectic SnPb solder. Pb-free soldering on Pb-free
HASL, OSP, immersion silver, immersion tin and immersion gold resulted
in the formation of a Cu-Sn IMC, which reduced the underlying pad and
through-hole copper thickness. With no barrier layer, any further thermal
exposure would continue to dissolve copper into the solder joint, forming
more IMC. Such dissolution was of particular concern where the initial
copper was thin, such as controlled impedance applications. Lead-free
wave soldering could be very aggressive, particularly during rework soldering, causing drastic reductions in copper thickness, most notably at the
through-hole entrance or "knee".

Voiding was another example of an issue that affected soldering on all
board surface finishes to some extent. The specific issue of microvoiding
with respect to immersion silver was examined with more work needed in
this area. It was important to note that IPC design standards had not

currently changed in response to Pb-free requirements. Land patterns in the newly released IPC-7351A met requirements for surface-mount assembly with either SnPb or Pb-free solders. Selection of the proper PCB surface finish would be more critical with increasing demands, particularly in light of the continuing drive to reduce costs. As a result, each finish needed to be examined and evaluated based on its recognized functional capabilities and known potential weaknesses.

The lead-free standards chapter reviewed some of the most common standards, which had been affected by the transition to lead-free soldering. Critical issues such as temperatures for lead-free surface mount soldering and visual inspection of lead-free soldered joints had been addressed to a certain extent but there were many more standards that need updating and this was being done as more data was becoming available.

Index